Propliners of the World
Volume 1

Propliners of the World Volume 1

Gerry Manning

FLIGHT
RECORDER
PUBLICATIONS

A passion for accuracy

First published in Great Britain in 2011 by
Crécy Publishing

© Gerry Manning

ISBN 9 780955 426841

Printed in Malta by Gutenberg Press Ltd

Crécy Publishing Limited
1a Ringway Trading Estate
Shadowmoss Road
Manchester M22 4LH
www.crecy.co.uk

CONTENTS

INTRODUCTION

In preparing this work the first task was to go through my slide collection to see what propliners I had photographed over the years. The number pulled out in the first instance was enough for several volumes, but on closer examination there were types no longer to be seen in the air, operators that had long ceased trading, and locations that no longer welcome nor host round-engine types. Sadly, the world of pounding pistons is in a decline and most of the operators are found in too few places. These classics have moved to some of the remotest places in the world, where they are simply irreplaceable.

Should the reader feel the urge to travel to Yellowknife, Fairbanks or Villavicencio, then do so because before too many years have passed the wonderful sound, smoke and smell of a piston engine bursting into life, on an operational aircraft, may have vanished for ever.

All the pictures in the book are my own from travels to many places over the years.

For those who would like to keep up to date with the subject, the excellent quarterly magazine *Propliner* is a must. For details visit www.propliner.co.uk

The basic premise of this book (in two volumes) is to look at some of the great classic propeller-driven aircraft as they operate around the world, sadly in ever-decreasing numbers. Each volume has four main sections.

In Volume 1 they are:

The Douglas DC-3
Bush and floatplane flying
Preserved aircraft and pleasure flying
First-generation turboprops

In Volume 2 they are:

Water-bombers
Russian and Chinese-built aircraft
South American operations
Cargo flights

Some aircraft, such as the DC-3, will appear in a number of sections. Of course the DC-3 deserves a section of its own, but also offers passenger flights, it still hauls cargo, and can be found in South America.

1   The Douglas DC-3

On 17 December 1935, when the first Douglas DC-3 took to the air from the company's Santa Monica, California, factory, it was unlikely that anyone who witnessed that first take-off could or would have predicted that more than seventy-five years later not only would examples be still operating revenue-earning flights in both passenger and freight configuration, but that the aircraft would have outlasted most of the so-called DC-3 replacements. These were designs of the late 1950s and early 1960s such as the Handley Page HPR-7 Dart Herald, Avro 748, NAMC YS-11 and Fokker F.27 Friendship, which are mostly now operated in smaller numbers or, in the case of the Dart Herald, have been completely retired.

What was it that led to such longevity? It was the Second World War and the military variant, the C-47 (known in RAF and RCAF service as the Dakota), being produced in such large numbers that, following the end of the conflict in 1945, many were available to the airlines of the world at war-surplus prices. It was also an aircraft that was able to be maintained by engineers world-wide and, not being pressurised, the airframe and components had no time-life restrictions. Added to this, all parts were replaceable and, even following accidents and crash damage that would write-off most airliners today, the DC-3 was able to be repaired and fly again.

The only dark clouds on the horizon to prevent the aircraft flying for ever are the lack of engine spare parts and the availability of aviation gasoline in some parts of the world. To counteract this, in both the USA and South Africa there is an active conversion programme to replace the piston engines with turbine power, so the machine that was voted by *Fortune Magazine* in 1999 as the 'greatest invention of the century' in the travel category will fly on for many more years to come.

Top: *Since its first flight in December 1935 the Douglas DC-3 has proved to be one of the few aircraft of which it can be said, without fear of contradiction, that it has changed the world. If on that day in Santa Monica, California, somebody had said that the design would still be in scheduled revenue-earning passenger service more than seventy-five years later, they would have been thought a madman. But to this day Canadian carrier Buffalo Airways, based at Yellowknife in the remote Northwest Territories, flies a regular passenger service across the Great Slave Lake to Hay River. The company has a fleet of nine DC-3s, some cargo-carriers and three with twenty-seven passenger seats; these, however, can be removed if cargo space is required. Pictured at Yellowknife is* **Douglas C-47 C-GWZS** *(c/n 12327), one of the passenger fleet.*

Middle: *It was during the Second World War and the need for hundreds of transport aircraft that the DC-3 was found to be the right design for the time. It had entered service with the United States Army Air Corps (which became the United States Army Air Force during 1941 and was not an independent service until the formation of the current United States Air Force in September 1947). The type had a number of different military designations, but the most produced was the C-47. This was different from the civilian DC-3 by the fitting of a large cargo door on the port side of the rear fuselage. Showing this to good effect is Buffalo Airways* **Douglas C-47 C-FLFR** *(c/n 13155) at the company's Yellowknife base.*

Bottom: *Built as a civil airliner in February 1941 for Eastern Airlines, prior to the United States' involvement in the Second World War,* **Douglas DC-3 N79MA** *(c/n 4089) can still be found in service to this day. Having passed through a number of owners and operators and registrations, it is now operated by Missionair, whose role is to fly medical supplies, clothing, blankets, clean water and food to disaster areas. Its area of operations is the Caribbean and Central America. As the company's name suggests, there is a strong Christian mission part to its operation, with all staff being volunteers, be they aircrew, engineers, doctors, nurses or other specialists. The company is based at Orlando-Kissimmee in Florida, where the aircraft was pictured.*

Above: *The largest state in the USA is Alaska, and without air transport many communities could not exist. They rely on aircraft to deliver everything they need. As well as the obvious items, one other is vital, this being fuel for generators to provide light and heat. This can be either petrol or diesel, and it is carried by specialist companies with their aircraft fitted with tanks in the fuselage. One such was Woods Air Service, based in Palmer, Alaska. Pictured at base is* **Douglas C-47 N50CM** *(c/n 13445). Built in 1944, it served with the USAAF and joined the American civil register in 1946; over the following years it has changed owners on many occasions, being registered in Mexico for some years before rejoining the US register in 1972. It is pictured with the snow-capped Chugach Mountain range in the background.*

Top: *Another of Alaska's ad hoc freight carriers is Bush Air Cargo, based at Anchorage International Airport. The company advertises with the slogan 'No runway? – No problem', and will haul up to 7,500lb (3,400kg) in the summer on wheels and 5,500lb (2,500kg) in the winter with the aircraft fitted with skis. The aircraft has 1,200cu ft of space with a range of up to 1,200 miles (1,930km) to any location in the state. If it will fit through the 7-foot (2.13m) cargo door, Bush will carry it. Pictured at Palmer, Alaska, is* **Douglas C-47 N777YA** *(c/n 25634) with company titles. This aircraft, built in 1944, first served with the US Navy with the designation R4D-6 before taking up a civil identity in 1946. It has spent most of its career in the largest state of the union with several owners and operators.*

Top: *In the United Kingdom the DC-3 is more commonly known as the Dakota, and the last large user was Coventry-based Air Atlantique/Atlantic Airways. The company had three main roles for the aeroplanes: passenger charters, ad hoc freight work, and pollution control. Oil spills around the coast of any nation are a problem for fishing and tourism as well as the environment in general. To combat these some of the company aircraft were fitted with spray bars under the tailplane to discharge dispersants. Pictured on take-off from its Coventry base is* **Douglas C-47 G-AMSV** *(c/n 32820) with full 'Pollution Control' titles. The red spray bar is visible under the rear fuselage and there is also a turbine to blow the chemical across the area being sprayed.*

Above: *Pictured on take-off from its Coventry base is Air Atlantique* **Douglas C-47 G-AMRA** *(c/n 26735), one of the company's freight fleet and available at short notice for such charters as car parts. A motor industry assembly line costs many thousands of pounds per hour to run; should a part not arrive on schedule, the line might need to stop at vast cost to the company. When this looks to be a possibility the supplier may charter an aircraft to deliver the missing parts. Although this can be expensive, it is far cheaper than allowing the line to stop and risking a penalty payment for a breach of contract to supply.*

Top: *In today's financial world it can be very expensive to use a new aircraft as a test-bed and not fly it a great deal. So, with the relatively low acquisition costs of a DC-3 as well as low running costs, it is no surprise that Racal Electronics Company leased a Dakota from Air Atlantique for trials of the radar for the BAe Systems Nimrod. Pictured on take-off from the company base of Coventry is* **Douglas C-47 G-ANAF** *(c/n 33436) with Racal titles and the radar being tested in a large housing under the nose. Built in 1945, this aircraft served with the Royal Air Force before joining the British civil register in 1953. It has been operated for most of its life except for a few years on the American register.*

Bellow: *The state of Florida is one of the most popular holiday destinations in that vast country. However, if it was not for the work of the DC-3 it would be an intolerable place to visit due to an infestation of mosquitoes. The hot climate and large areas of standing water are ideal breeding grounds for these annoying insects. In areas of highest risk the local county authorities will use aircraft to spray the centres of breeding to kill off the mosquitoes.* **Douglas C-47 N10005** *(c/n 25527) carries 'Mosquito Control' titles from its days with Hillsborough County Health Unit when based at Tampa. It is seen at Lehigh Acres, Florida.*

Top: *Lee County Mosquito Control District covers the town of Fort Myers on Florida's west coast. Its fleet of DC-3s carry no titles and are based at Lehigh Acres-Buckingham Airpark. Pictured on the ramp at base is* **Douglas C-47 N836M** *(c/n 25977). This airframe dates back to 1944 and wore an American military uniform for most of its long life, first with the Air Force then in its last few years with the US Army. It was not until the mid-1970s that, after a couple of years in store, it joined its present operator.*

Above: *For fans of real propeller-driven aircraft the sight of seven Douglas DC-3s lined up in the sun and all in wonderful condition is a joy to behold. Pictured at Lehigh Acres-Buckingham is the fleet of Lee County Mosquito Control District. Normal spraying usually takes place at first light or dusk, when the air is at its stillest and the spray will just cover the required areas of calm flat water and not affect other places.*

Top left: *The town of Naples can be found on the southern end of the west coast of Florida. Based here is Collier Mosquito Control District, and seen on the ramp is **Douglas C-47 N842MB** (c/n 19741). A 1944-built airframe, it served with the US military until 1965 and took up its current role in 1967 when it joined the US civil register.*

Bottom left: *Also pictured at Naples, Florida, are four DC-3s operated by the Collier Mosquito Control District. As all good things must come to an end, the cost of fuel and its availability for the DC-3 have seen the county start operations with turboprop-powered Short Skyvans.*

Above: *Nord Aviation was founded in 1955 and operated from Pembina, North Dakota, before moving its operating base to Santa Teresa, New Mexico. Its role is a general air cargo carrier with an all-propliner fleet. Pictured at base is **Douglas C-47 N57626** (c/n 4564). This 1942-built aircraft was disposed of following the war's end and, despite having a number of owners over the following years, has always kept the same registration.*

Below: *Photographed on the ramp at Gimli, Manitoba, is **Douglas C-47 CF-OOW** (c/n 13342) in the livery of Enterprise Airlines. The main role of this carrier was to feed the just-in-time delivery of parts to the motor industry. Based in Oshawa, Ontario, the company also operated a flying school and fixed base operation for visiting aircraft. This DC-3 dates from 1944, and served with the Royal Canadian Air Force before being demobbed in 1971, since when it has had a number of operators. It is of note that the registration letters still begin 'CF-' (the old-style Canadian markings) when the correct version should be 'C-F...'.*

Top: *Pictured at California City, California, is* **Douglas C-47 N193DP** *(c/n 4433). The concept of the private aircraft is common in the USA, the size and vintage being dependent upon the thickness of the owner's wallet. This 1942-built example is one such aircraft. Its history began with service in the US Marine Corps before entering the civil market in 1946. Over the years it has had roles in North and South America as a corporate transport and as a flying survey aircraft.*

Above: *For many people the thought of jumping out of a perfectly serviceable aeroplane fills them with horror, yet for others it is the great sport of sky-diving. These are people who have many more take-offs in aircraft than landings. During the Second World War massed parachute jumps took place at locations such as Normandy on D-Day and in Holland at Arnhem, and the aircraft used was of course the C-47 (DC-3). What worked for the military would also do the job for the sports jumper, as a lot of people could be packed into the fuselage and this would keep down the cost of each jump. The western states of America, with their excellent weather, host many centres of the sport. Pictured at Lodi in central California is* **Douglas C-47 N4991E** *(c/n 12106), used by the local centre as a jump platform.*

Opposite top: *In each of the recent years at Oshkosh, Wisconsin, during the EAA (Experimental Aircraft Association) fly-in,* **Douglas C-47 N1XP** *(c/n 4733) has been used to drop the Liberty Parachute Team. The bright yellow aircraft is named 'Duggy' and has titles proclaiming it as 'The Smile in the Sky', as a line painted under the cockpit gives a 'smile' effect. The airframe was built in 1942 and spent its war years with the US military in Australia. It returned to the USA in 1946 and has been operated and owned in Canada as well as mainland USA*

Opposite bottom: *Regulation of United States aviation is under the jurisdiction of the FAA (Federal Aviation Authority) and, as would be expected, it operates a fleet of aircraft from its base in Oklahoma City. Still on its books is* **Douglas C-47 N34** *(c/n 33359). Its role today is to act as a flying educational display covering the various aspects of FAA operations, and is pictured here at Oshkosh, Wisconsin. A 1945-built aircraft, it served for the first ten years of its life with the US Navy before being stored in 1956. The following year it joined the fleet of the CAA (Civil Aeronautics Authority), the forerunner of the FAA.*

Left top: *Herpa Wings is a German model manufacturer founded in 1949 and based at Dietenhofen. To advertise its products there could be no finer display billboard than a classic aircraft.* **Douglas DC-3 N142D** *(c/n 2954) is pictured at Oshkosh, Wisconsin, with company titles and with an extra logo on the fin celebrating the type's seventy-fifth anniversary. This airframe is a genuine pre-war example, having been built in 1938 and serving with Swissair until the middle of the 1950s, when it was replaced by Convair 440s. It then began its American career with Ozark Airlines for a decade before being sold on to a number of subsequent operators.*

Left middle: *Usually, when an aircraft is acquired by a museum it has performed its last flight. Pictured at Oshkosh, Wisconsin, is one of the exceptions to this,* **Douglas C-47 N97H** *(c/n 33613) of the Hiller Museum in San Carlos, California. Hiller is of course a well-known helicopter manufacturer, and the museum does lead on these, but the DC-3 acts as a flying advertisement when it attends air shows in the western states of the USA. Built in 1945, it only had a short war career before entering the civil market the following year. For much of its life it has acted as a corporate aircraft, taking business executives to their meetings around the USA.*

Left bottom: *Also pictured at Oshkosh, Wisconsin, is* **Douglas C-47 N728G** *(c/n 4359) in the house colours of the manufacturer. Owned by an Alabama-based company, it has a corporate interior, a role it has played most of its life. Originally built in 1942, it spent the war years in service with Texas-based Braniff Airlines as the carrier continued operations in support of the war effort.*

Above: *There is a growing trend in the world to paint aircraft in liveries of a bygone era. This ranges from the world's premier carriers putting new Airbus and Boeing airliners into what are referred to as 'retro' schemes – those with which the airline operated during the 1950s or '60s – to painting classic preserved aircraft in the colours of airlines that once flew them but are no longer operational. Into the latter category falls this* **Douglas C-47 N44V** *(c/n 4545) in the colours of Piedmont Airlines. The carrier was formed in 1948 and based in North Carolina; in 1989 it became US Air. This airframe never actually served with Piedmont, but its owner, the Carolinas Historic Aviation Museum, has adopted a long-gone local-based company as the basis for its aircraft's colours. It is seen at Oshkosh, Wisconsin.*

Top left: **Douglas DC-3 NC17334** (c/n 1920) is in the colours of American Airlines as 'Flagship Detroit', the name of the foundation that owns this aircraft, one of the oldest DC-3s still flying. It is based at Bomar Field, Shelbyville, Tennessee, and first flew in 1937 operating with American Airlines in the livery it now has. The current operator took over the aircraft in 2005, and it is pictured flying at Oshkosh, Wisconsin.

Bottom left: Based at Waxahachie, Texas **Douglas C-47 N737H** (c/n 6062) first served with the US Navy in 1942, was de-mobbed in 1946 and then served with a number of American industrial companies in various roles joining its current owners, Airborne Imaging, in 2002. As its name suggests their role is to survey and map areas by means of an assortment of sensors that can be fitted to its belly. In its striking white and red colour scheme it is pictured landing at Oshkosh, Wisconsin.

Top: An aircraft with one of the highest flying times of any type is **Douglas DC-3 N18121** (c/n 1997). This 1937-built example has more than 91,000 hours in the air – more than ten years airborne.

It started its life with Eastern Airlines and has returned to its original colours of the 1930s-style 'The Great Silver Fleet', with those titles on the cabin roof and the name of the airline in the logo behind the cockpit. This picture at Oshkosh, Wisconsin, shows the passenger door open with its built-in air stair, while aft of this the baggage hold door is also open. Privately owned, the aeroplane is based at Troutdale Airport in the north-western state of Oregon.

Bottom: In Western Airlines period colours is **Douglas DC-3 NC33644** (c/n 4123). Built in 1941, before America joined the war, it first served with United Airlines before joining the fleet of Western the following year. That company operated it until 1958 when it started a career with a number of owners. The current owner is a retired airline captain who keeps it on his home airstrip at Oakville, Washington State. Western Airlines had a history dating back to 1925, making it one of the oldest airlines in the USA. In April 1987 it was taken over and incorporated into Atlanta-based Delta Air Lines, and the long-established name was lost. It is pictured at Oshkosh, Wisconsin.

Top: *The Peruvian Navy (Armada Peruana) had two main uses for its fleet of C-47s: one was the normal cargo-carrying task, and the other was that of coastal patrol along the nation's Pacific shore. Pictured at Lima is* **Douglas C-47 AT521** *with naval titles.*

Above: *Long after air forces began flying jets in the strike and fighter roles, the transports, especially in the developing world, still had piston engines and the favourite was the ever-reliable design from Douglas in its most common military variant, the C-47. Once such nation was Colombia, where the Fuerza Aerea Colombiana operating many C-47s and to this day flies the turbine-engine conversions. Pictured at Madrid Air Force Base, Colombia, is* **Douglas C-47 FAC 1670** *being operated in the cargo role.*

Top: *Built in 1942 and operated by the US Army Air Force in Australia, **Douglas C-47 VH-AES** (c/n 6021) is pictured at Avalon near Melbourne and has stayed in the land down under all its life. It was eventually withdrawn from use in 1973 and preserved at Melbourne's Tullamarine Airport. However, since it is impossible to keep a good Dakota down, September 1988 saw it take to the air once more following a full restoration and return to its 1946 livery of Trans Australian Airlines with the name 'Hawdon'. It is of note that with many immediate post-war colour schemes very little actual paint is used, with most of the aircraft being highly polished metal.*

Bottom: *One of the smartest examples of the type flying today is **Douglas C-47 N7AP** (c/n 25341) in the livery of the University of Ohio based at Athens. Pictured at Oshkosh, Wisconsin, it was built in 1944 and operated by the US military. Following the end of the war it served with a number of branches of the federal government, including the CAA, FAA and the Department of Transport, before taking up its current academic career in 1981. Its uses have included the collection of data from electro-magnetic signals from the university's Avionics Engineering Centre.*

Top *Pictured at Oshkosh, Wisconsin, **Douglas C-47 N103NA** (c/n 9531) carries the titles 'Flabob Express', a reference to its owner, Flabob Aviation Associates, and its base at California's Flabob Airport. Built in 1943, it joined the Royal Air Force and served in a communications role flying the famous General (later Field Marshal) Claude Auchinleck in India. Following the partition of the sub-continent in 1947 it joined the Pakistan Air Force for the next five years before being sold in the USA and later Canada.*

Above: *The 2010 EAA Fly-in at Oshkosh, Wisconsin, featured the 75th Anniversary of the DC-3. Perhaps the most remarkable of the attending aircraft was **Douglas C-47 N74589** (c/n 9926), as just seven weeks before it arrived at the air show it had been derelict and blocked in by trees that had grown up since it had been parked. The bulk of the work was done by two engineers working seven days a week with very long hours. Eventually, when it was ready, a week before the event, the first test flight found a serious engine problem and a new powerplant had to be purchased, delivered and fitted. The job was more than a quick engine swap, as the oil lines and tanks had to be cleaned out as well. This new unit did the trick and the aircraft flew from Covington, Georgia, all the way to Wisconsin to be the last of the arrivals but, for the weary crew, the most satisfying.*

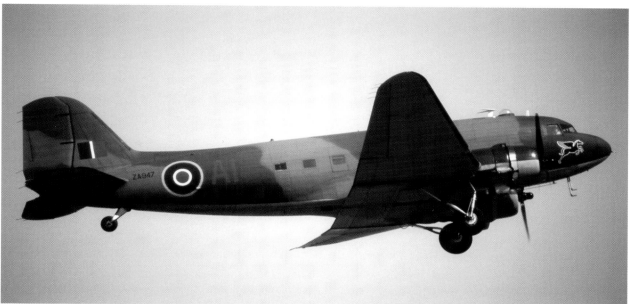

Top: *The Royal Air Force today still has a Dakota on its books, for two very different roles.* **Douglas Dakota 3 ZA947** *(c/n 10200) is operated by the Battle of Britain Memorial Flight as an historic aircraft type in its own right, and one that played an important role in the Second World War, with one pilot, Flight Lieutenant David Lord, winning the Victoria Cross during the Battle of Arnhem in September 1944. Pictured at Berlin-Schönefeld, the aircraft has both of its Pratt & Whitney R-1830 radial piston engines turning. Visible on the nose are the 'Pegasus' markings of 267 Squadron, in whose colours it currently flies.*

Above: *The other role of the BBMF's Dakota is that of a pilot trainer for the unit's Avro Lancaster. In today's RAF no pilots have experience of four-engine piston-powered aircraft with tail wheels; the last*

similar type was the Avro Shackleton, which was retired in 1991. To teach the skills needed to fly this important historical artefact, only one other Lancaster is airworthy and that is in Canada, so pilots learn multi-engine tail-wheel operations on the Dakota, first as co-pilot then as captain. Following this they can move on to the Lancaster as co-pilot before changing to the left-hand seat. This also keeps the current Lancaster crew in relevant flying experience as hours on such a valuable type are kept as low as possible per year to prolong its life. **Douglas Dakota 3 ZA947** *(c/n 10200) is pictured taking off from Berlin-Schönefeld. The history of this aircraft dates back to 1943, and it spent its life in the Royal Canadian Air Force before arriving in the UK in 1969, when it joined the Royal Aircraft Establishment, moving to the BBMF in 1993.*

Top: *As has been seen earlier, many owners of private DC-3s have put them into airline liveries in which they once operated. The other view is to put your aircraft into one of the schemes it carried when in military service. French-owned 1943-built* **Douglas C-47 F-AZTE** *(c/n 9172) wears a false Royal Air Force serial of KJ994 when pictured at Cazaux in south-west France; that aircraft had been used by Lord Tedder when he was Chief of the Air Staff, and was scrapped in 1963. The real history of F-AZTE was from 1952 to 1971 with the French Air Force (Armée de l'Air) before a civil career with a number of operators in France.*

Above: *It would be expected that the Commemorative Air Force (formally known as the Confederate Air Force) would fly the aircraft it owns in military livery. Pictured at Willow Run, Detroit, is* **Douglas C-47 N47HL** *(c/n 27203) in the USAAC pre-Second World War markings, being largely polished metal with a three-colour (red/white/blue) rudder. This Texas-based aircraft never in fact operated with the US military, but when built in 1945 joined the RCAF and was operated by it until 1972, when it was demobbed to the civil market, first in Canada then the US.*

Top: *Operated by the Valiant Air Command based at Titusville, Florida, **Douglas C-47 N3239T** (c/n 19054) is pictured at Oshkosh, Wisconsin, wearing its correct original USAAF serial of 42-100591 – the '42' prefix relates to the fiscal year it was ordered, although it first flew in 1943. The colour scheme is that in which it flew during D-Day, complete with black and white invasion stripes. Following US service, the aircraft operated in Scandinavia, first with the Royal Norwegian Air Force from 1950 to 1956, then the Royal Danish Air Force, with which it operated until 1983 when the current owners took it over.*

Bottom: *Pictured at Oshkosh, Wisconsin, is **Douglas C-47 N8704** (c/n 33048), owned and flown by the Yankee Air Force, based at Willow Run, Detroit. The colour scheme in which it operates is that used by the USAF during the 1950s and '60s. The serial it carries, '476716', is its correct original military markings, with just the first '4' from the 1944 fiscal missing.*

Top: *It is not unusual to see vintage Second World War aircraft in what are 'bogus' schemes. What the owners are trying to do is to recreate and honour either complete units or individual aircraft. One such is* **Douglas C-47 N99FS** *(c/n 12425), owned by Mr Don Brooks and based at Griffin, Georgia. It wears the markings of the 'Jungle Skippers'; these were four squadrons that made up the 317th Troop Carrier Group, part of the 5th Air Force operating in the South Pacific area. The name originated from a war correspondent, who wrote how he had flown on a 'jungle-skipping' aircraft; the unit's commander liked the name, so it started to appear on its aircraft. The aircraft's real history was that it served in the RCAF, and it is pictured here at Oshkosh, Wisconsin.*

Above: **Douglas C-47 N150D** *(c/n 4463) is seen at Oshkosh, Wisconsin, in a hybrid scheme. This 1942-built aircraft spent its war years flying for Pan American Airways based in Africa. Following French military service, it operated for the Israeli Defence Force/Air Force, in whose basic camouflage scheme it now operates. The owner is the Ozark Airlines Museum, based at St Charles, Missouri. Ozark flew a large fleet of DC-3s from 1950 until the late 1960s, and was taken over by Trans World Airlines in 1986. In addition to the basic colours, the aircraft carries its full original serial 41-18401 and the early Second World War USAAC roundel on the rear fuselage.*

Top: *Palmer, Alaska, is the setting for views of three Douglas DC-3s as they await their next cargo shipments.*

Right: *Showing off her nose art is **Douglas C-47 N47HL** (c/n 27203). Titled 'Bluebonnet Belle', it is operated by the Highland Lakes Squadron of the Commemorative Air Force based at Burnet, Texas, the Bluebonnet capital of the state. It frames the Canadian Warplane Heritage Museum's 1939-built DC-3 C-GDAK (c/n 2141) at Oshkosh, Wisconsin.*

Top: *An Oshkosh nose and tail shot, but the registration 'NC43XX' is false. This aircraft, operated by Thunderbird Flying Services, is in fact* **Douglas C-47 N353MM** *(c/n 11665), but the period markings look so much better.*

Left: *The nose view of a pair of Douglas DC-3s at Oshkosh could not be mistaken for any other aircraft, so well known is the classic machine's profile.*

Top right: *A sight that may never be able to be repeated in the United Kingdom – seven Douglas DC-3s in formation. They are the fleet of Air Atlantique and are pictured at the company's Coventry base.*

Bottom: *The search for a 'DC-3 Replacement' has foxed many aircraft manufactures, including Douglas itself. The termination of production orders following the Second World War saw the company naturally looking to build a new airliner for the hoped post-war boom in passenger flying. Rather than embark upon a completely new design, Douglas decided to modernise the DC-3 airframe. The fuselage was lengthened by 3ft 3in (0.99m) to add to the seating capacity, a new larger vertical tail was added, and the horizontal tailplane and wing tips were squared off. The engine nacelles were redesigned to make the retracted undercarriage fully enclosed, and finally new powerplants in the form of a pair of Wright Cyclone R-1820 air-cooled radial piston engines of 1,475hp were fitted. Now known as the DC-3S, or Super Dakota, it took to the air for the first time on 23 June 1949 from Clover Field. Sales, however, amounted to just three aircraft, ordered by Capital Airways. As with most aircraft from the Douglas company, there was nothing wrong with the new one; its problem was the vast number of war-surplus C-47s on the market at a fraction of the price of new-built designs. Salvation came with an order for 100 from the US Navy under the designation R4D-8 (C-117 following the 1962 standardisation of US military types). These aircraft served both the Navy and the US Marine Corps into the mid-1970s before being sold on to the civil market. Pictured at its Anchorage, Alaska, base is* **Douglas C-117 N851M** *(c/n 43302) of Trans Northern, configured for cargo operations.*

Top left: *Also seen at its Anchorage, Alaska, base is Trans Northern's cargo-configured* **Douglas C-117D N28TN** *(c/n 43354). Once flown by both the US Navy and USMC from bases as far apart as Formosa (Taiwan) and the UK, it has spent its civil life mainly in Canada, where it wore its current highly visible paint scheme. This would be very useful if the aircraft had to make a forced landing in the wilds of Alaska, as it would be seen for miles.*

Middle left: *Nord Aviation, based at Santa Teresa, New Mexico, operates a single* **Douglas C-117D N321L** *(c/n 43345) alongside two standard C-47s for general cargo operations. The aircraft is pictured at base.*

Borrom left: *The Colombian Air Force has been an operator of at least two C-117s following their US Navy service. Pictured at Madrid Air Force Base is* **Douglas C-117 FAC 1632** *(c/n 43360). This airframe joined the USN in 1952 and served for twenty years, followed by another ten in store at Davis-Monthan AFB in Arizona before being sold on.*

Below: *The biggest threat to the DC-3 flying on for ever is not some new and shiny aircraft with amazing performance and low costs, but fuel, or rather the lack of it. The piston-powered machines* require aviation gasoline (AVGAS), and in many parts of the world it is simply not available. All jets and turboprops run on aviation turbine fuel (AVTUR), and this is the only fuel found at many of the world's major airports. Also the faithful R-1830 engine in the DC-3 is no longer being produced, so spares will eventually run out and the powerplant can have only so many overhauls before it needs replacement. The solution for the evergreen DC-3 has been turbine power. The first aircraft to be re-engined dates back to 1949, with Rolls-Royce Darts being fitted for trial purposes; following this, the original pistons were refitted. Today the two main centres for conversions are the USA and South Africa. Pictured at Ysterplaat, Cape Town, is* **Douglas DC-3-TP 6877** *(c/n 11925) of 35 Squadron of the South African Air Force. The roles undertaken include that of maritime patrol of the nation's vast coastline.*

Bottom: *Another view of* **Douglas DC-3-TP 6877** *at Ysterplaat shows the new powerplant configuration to advantage. All the new conversions have replaced the original P&W R-1830 air-cooled radial pistons with Pratt & Whitney Canada PT-6 turboprops driving a five-bladed propeller. Since these are much lighter than the originals, the centre of gravity has changed; to correct this, two plugs have been added to the fuselage length to increase its size and capacity.*

Top: *Pictured on the move at Oshkosh, Wisconsin, is Douglas DC-3-65TP ZS-MAP (c/n 32644). This South African conversion served that nation's military all its life and has now taken the US registration N882TP and is on the strength of the National Test Pilots School based at Mojave, California. The organisation is a civil-run establishment to teach the various aspects of testing aircraft from the standpoints of both pilots and engineers, from short courses on aspects of a single avionics package to year-long courses to qualify as a test pilot on both fixed and rotary-wing aircraft.*

Bottom: *California City, California, is the location of South African conversion Douglas DC-3-65TP N145RD (c/n 20175), then operated by Dodson International Air of Covington, Georgia. It is now with Baja Air, and was in fact the first of the South African conversions. The 1944-built airframe had spent much of its civil career in the Middle East before joining the South African Air Force as late as 1989.*

Top: *The leading converter of the DC-3 is Basler, based at Oshkosh, Wisconsin. With more than fifty airframes converted and plenty of stock available, the type should fly on for many more years. The new turbine engines deliver 1,642shp compared to the piston's 1,200hp, thus increasing the speed from 200 to 225 knots. The Colombian Air Force uses the conversion, one as a normal transport machine and the other as a 'gun-ship'. During the Vietnam War the flying gunship was developed first with the AC-47, then with the AC-119 and on to the current AC-130 Hercules. Using the newly re-engined DC-3 (C-47) as a gunship is a cost-effective means of dealing with the two problems facing the government in Bogota, the FARC left-wing rebels and the drug cartels, both of which operate in remote jungle areas. Pictured at Gomez AFB, Apiay, is* **Basler AC-47T 'Fantasma' FAC 1681** *(c/n 33248); the 'Fantasma' name is a locally applied Colombian one.*

Bottom: *The Douglas DC-2 was a stretched version of the single DC-1 produced. It was ordered by TWA and first flew from Santa Monica in May 1934. Although a successful aircraft, it was deemed to be underpowered with the original pair of 750hp Wright SGR-1820 radial pistons, as well as being too small for economic loads. Pictured at Oshkosh, Wisconsin, in the period markings of TWA as NC13711 is* **Douglas DC-2-118B N1934D** *(c/n 1368). It is owned by Boeing's Museum of Flight in Seattle and is being flown by aviation legend Clay Lacy.*

*Pictured in the sky over Coventry is **Douglas DC-2-142 NC39165** (c/n 1404), a former US Navy machine in the colour scheme worn by KLM (Royal Dutch Airlines) DC-2 PH-AJU when it flew in the MacRobertson Australia Air Race. This ran from Mildenhall in the UK to Melbourne in Australia and was part of that city's 1934 centenary celebrations. The 11,300-mile (18,200km) course was won by the de Havilland DH.88 Comet Racer, a two-seater racing aircraft designed for the race, but in second place came the DC-2 carrying three passengers, the crew and a consignment of mail showing that the latest airliner could travel very long distances in a short time. The real PH-AJU, named 'Uiver', was destroyed in a crash in Iraq in December of that year.*

Bush and floatplane flying

It is a strange fact that Juneau, the state capital of Alaska, has no road access to the biggest two cities of Anchorage and Fairbanks – the only way to arrive is by sea or air. This is also true of many other settlements and towns in the largest state in America, and to huge swathes of Canada. Since these places are small and have few people, they do not have big airports with long all-weather runways and the latest navigation or landing aids. However, the natural features that can be found in abundance are freshwater lakes and areas of open ground, and from these circumstances has developed a type of aircraft to service such places – an aircraft that can land on rough unprepared strips or be fitted with floats to land on lakes during the long days of summer and, when equipped with skis, to land on either the snow-covered land or frozen lake in winter.

De Havilland Canada is the master at producing aircraft for bush or lake. Its two great designs are the DHC-2 Beaver and its larger development, the DHC-3 Otter. Both originally piston-powered, many airframes have been brought up to date by the fitting of a new, fuel-efficient, lighter and yet more powerful turboprop. Both designs had STOL (Short Take-Off and Landing) ability designed into them, so they are able to land and take-off in the shortest of spaces. Some have also been fitted with outsize very-low-pressure tyres, allowing a landing upon the roughest of terrain.

When fitted with floats for the thousands of lakes, they are also the driving force for a whole tourist industry – fly-in fishing. From small-town seaplane bases, fishermen and hunters are taken to remote lakeside lodges to stay for a number of days in areas of total wilderness yet with home comforts in the lodges. Locations such as these can be hundreds of miles from the nearest road.

Two of the largest cities of the US and Canada, Seattle and Vancouver, have floatplane operations from city-centre locations. These are ideal for the business traveller who needs to get to places that do not have airports yet can be reached from an ideal downtown site without all the hassle so prevalent at the world's large passenger airports.

Top: *Basking in its own reflection is **de Havilland Canada DHC-2 Beaver N6LU** (c/n 908) of Trail Ridge Air at the company base, Lake Hood Sea Plane Base, Anchorage, Alaska. The Beaver has proven to be the ultimate 'bush' aircraft to operate in the backwoods area. It has been fitted with floats for the many lakes that abound, skis for the long winter operations and either normal wheels or large 'tundra' tyres with very low pressure – as low as 7psi (pounds per square inch) – for unprepared ground. Trail Ridge uses its aircraft for a number of roles including charter flights into the Alaskan outback in the six-seat passenger craft, air taxi operations, and tourist sightseeing for the glaciers and mountains of America's largest state.*

Bottom: *The Beaver was first flown in August 1947 from Downsville, Ontario. It can haul up to 1,350lb (613kg) for 470 miles (752km) with a still-air take-off in just 595 feet (220m). Originally considered with a 330hp Gipsy Queen engine, this was replaced by a 450hp Pratt & Whitney Wasp Junior air-cooled radial piston motor, giving the aircraft its excellent STOL (Short Take-Off and Landing) performance. Pictured on a landing approach to Vancouver SPB (Sea Plane Base) is **DHC-2 Beaver C-GFDI** (c/n 606) of Baxter Aviation. This company was founded in 1985 and operated scheduled passenger services from its base at Nanaimo, Vancouver Island, to both Vancouver International Airport's SPB and to the city centre's floatplane terminal. It now flies under the West Coast Air name following a takeover.*

Top: *Pictured at La Ronge, Saskatchewan, is **DHC-2 Beaver C-GAEB** (c/n 703) of Prince Albert- based Transwest Air. This airline is the largest and longest established in the Canadian province, with scheduled passenger and freight services and charter services to a number of locations, using both fixed and rotary-wing aircraft. Lake Lynn in Manitoba is the site of its floatplane operations, and this aircraft is seen on its floats resting on a trolley, being prepared for winter storage.*

Bottom: *Vancouver Island Air is based at Campbell River Sea Plane Base, British Columbia. All its fleet are operated on floats, and pictured at base is **DHC-2 Beaver C-FWCA** (c/n 1285). Since 1985 the company has been operating charters and scheduled flights taking groups to locations that have no roads and can only be reached by water. When the Beaver was being designed a great deal of input was made by OPAS (Ontario Provincial Air Service); this local government unit made a number of suggestions as to what the ideal bush aircraft would need to do to be a success, including such seemingly simple needs as a door big enough to take a standard-size fuel drum.*

Top: *As well as the main fly-in at Oshkosh, nearby Lake Winnebago hosts a splash-in for the many floatplanes that attend this major aviation event. Pictured on the lake is **DHC-2 Beaver N101CB** (c/n 1398), a privately owned example. The largest users of the Beaver were the US Army and USAF, which operated nearly 1,000 airframes between them. When they were retired these well-maintained aircraft soon found homes in the remote places of the world where the lack of a long concrete runway was no barrier to flying services.*

Bottom: *With its P&W R-985 air-cooled radial piston engine turning, **DHC-2 Beaver N104RL** (c/n 498) powers along Anchorage's Lake Hood SPB. The aircraft is operated by Rainbow Bay Resort based at Pedro Bay, Alaska, and its role is to fly people to the rustic lodges operated by the company as a base for a fishing or hunting holiday on the shore of Lake Lliamna, 180 miles from the nearest road. As well as fishing trips the company also offers trips to hunt and shoot both brown and grizzly bears.*

Top: *West Coast Air is based at Vancouver, British Columbia, and is an all-floatplane airline with many years of operations from downtown Vancouver across to Victoria Harbour on Vancouver Island, as well as other BC locations. Added to this are charter flights to see the scenic wonders of western Canada, which include orca killer whales in Haro Strait. Pictured at base is **DHC-2 Beaver C-GHMI** (c/n 1215) departing with a load of passengers.*

Bottom: *Once operated by the British Army Air Corps, **DHC-2 Beaver N985P** (c/n 1624) is pictured with its piston engine pounding as it taxis to the dock at Kenmore SPB in Washington State. It was used by the British military from 1966 until the end of the 1980s, when it was put into storage. Shipped to the USA, it was restored at Kenmore for civil operations. The airframes used by the AAC were built in Canada at the company's plant in Toronto, then shipped to the UK for assembly at the Hawker Siddeley (previously de Havilland) factory at Broughton, Chester.*

Top: *Operating since 1981 from Vancouver, British Columbia, Harbour Air's fleet is a mix of three types, all single-engine floatplanes. Pictured at downtown Vancouver SPB is **DHC-2 Beaver C-FOCN** (c/n 44). The carrier operates scheduled services from here to Victoria, Nanaimo, Maple Bay, and Butchard Gardens, all on Vancouver Island, as well as Bedwell Harbour and Ganges on the Gulf Islands. Charter flights are also available. Many passengers use the Victoria to Vancouver service as a daily commute to work, as the alternative ferry trip takes several hours.*

Bottom: *Pictured at its Lake Union location in downtown Seattle is **DHC-2 Beaver N6781L** (c/n 788) operated by Kenmore Air. This company operates domestic services to points in Washington State such as the San Juan Islands, a popular holiday destination in the Pacific North West; locations include Deer Harbor, Eastsound, Fisherman Bay, Friday Harbor, Roche Harbor, Rosario and West Sound. As well as the domestic operations, the company flies an international service to Victoria Harbour, Vancouver Island, on an all-year-round basis as well as summer (May to September) routes to Galliano, Salt Spring and South Pender Islands, located in the Gulf Island chain.*

Top: *Kenmore Air **DHC-2 Beaver N72355** (c/n 1164) is pictured at the company's Lake Union base, the engine turning as it heads into the dock. The power is usually cut prior to arriving, and the aircraft will have enough forward momentum to glide into position, then be manhandled to the dockside. As well as maintaining its own fleet, Kenmore offers maintenance and rebuilding services to Beaver owners, including engine overhauls, sheet metal repair, fitting and servicing floats, avionics, upholstery and painting. A derelict airframe can be taken in to emerge fully rebuilt to the highest standards with all the latest navigation equipment.*

Bottom: *One of Canada's main industries in the outback is tourism – it is a place to see nature at its finest in unspoiled and unpopulated areas. The pastimes here are fishing and hunting, and a whole network of air transport operators and specially constructed lodges at remote lakeside locations has sprung up to service this need. Pictured at Kakabeka SPB, Ontario, is **DHC-2 Beaver C-GUNE** (c/n 1403) of locally based Kakabeka Air Services.*

Top left: *Pictured at Kenmore Air Harbor, its North Lake, Washington, facility, are three DHC-2 Beavers out of the water. Aircraft will be taken out of the lake for maintenance inside the hangar, following which they will be parked outside to await collection by the owner.*

Middle left: *Kashabowie Outposts is another of the carriers catering for the wilderness tourist. From its Eva Lake location situated some 30 miles east of Atikokan, Ontario, it flies to ten lakeside lodges for fishing or moose-hunting. The aircraft will fly the passengers to lodges that will sleep up to ten people in rustic comfort, and which are usually provided with cooking stoves powered by propane gas and fridges to preserve the fish that are caught. No roads lead to these locations, so the guests will have a peaceful time until the sound of the aircraft drones overhead and it lands on the lake, heralding the end of their holiday. Pictured at Kashabowie River SPB is **DHC-2 Beaver C-FOCC** (c/n 23). Following a change in the rules by Transport Canada (the regulatory body), the company has recently been allowed to carry external loads on its Beavers; it can now haul such items as a 210lb canoe, 22ft x 40in x 30in deep, or a 270lb boat, 14ft 5in x 5ft 2in x 2ft 6in, as well as other similar items. The loads are all fixed to the float and its support struts by rope.*

Bottom left: *How to make a good aircraft even better – fit a new engine and extend the fuselage length. One of the problems with piston-powered aircraft is that they use a lot of fuel, and that is AVGAS (aviation gasoline), which is getting more and more expensive and its supply is not always readily available in some locations. The answer is to re-engine with a Pratt & Whitney Canada PT-6A 550shp turboprop. Since this powerplant is much*

*lighter than the Wasp Junior, to maintain the centre of gravity the fuselage is extended by 5 feet, meaning that two extra revenue-paying passengers can be carried as well as an extra fuel tank. Pictured at Lakeland, Florida, is **DHC-2 Turbo Beaver Mk 3 N388N** (c/n 1642TB23) owned by a Florida-based holding company.*

Top: *Yellowknife, Northwest Territories, is the home of Arctic Sunwest. Established in 1993, it operates a mixed fleet from both Yellowknife Airport on wheels in summer and skis in winter, and from the seaplane base on the shore of Great Slave Lake in Yellowknife's old town. Pictured at its airport base is **DHC-2 Turbo Beaver Mk 3 C-FOPE** (c/n 1691TB59); as can be seen in this side view, the fuselage extension is apparent, and the aircraft has a taller and more squared-off vertical tail fin. It is of note that within the floats are retractable wheels, making this a true amphibian, able to operate from both land and water.*

Bottom: *Pictured while in store at Kenmore Air Harbor, North Lake, Washington State, is **DHC-2 Turbo Beaver Mk 3 N14TB** (c/n 1638TB21). As well as the extra seating in the turbo version, it will carry more cargo, and climb to height and cruise faster than the piston-powered model. However, many companies will continue to use the original aircraft as the very high cost of converting to turbine power does not pay for itself when they are only used in the summer months and spend the rest of the year in store.*

Top: *Following the success of the Beaver in service, De Havilland Canada was soon asked for an aircraft with the same STOL performance but with twice the capacity. The name King Beaver was originally planned, but this was changed and the new aircraft was named the Otter. The powerplant for the new design was the 600hp Pratt & Whitney R-1340 Wasp air-cooled radial piston engine, and the first flight was in December 1951 from the company's Downsville, Ontario, site. Its go-anywhere, carry-anything ability soon found favour with military operators in both the US and Canada. In fact, to this day an Otter and a pair of Beavers serve with the US Navy's Test Pilot School, offering a type of flying few, if any, fast-jet pilots will have experienced. Pictured at Nestor Falls SPB, Ontario, is **DHC-3 Otter C-GYYS** (c/n 276) of Northwest Flying.*

Bottom: *Northern Wilderness Outfitters may sound like a shop for lumberjacks, but is in fact a Fort Francis-based company flying people to its fishing lodges in northern Ontario, which it has been doing for more than forty years. Pictured on Rainy Lake is **DHC-3 Otter C-GUTL** (c/n 365) with its P&W Wasp engine turning a three-bladed propeller. This aircraft was first built in 1960 and operated by the Royal Canadian Air Force. Its roles have included liaison and light transport, and for a period in 1965 it flew in Pakistan in support of the United Nations' mission supervising a cease-fire between Pakistan and India. The long flight from Canada to the sub-continent was easy, as it was dismantled and fitted into a C-130 Hercules. It was later put up for disposal and sold in 1982, taking up its Canadian civil identity.*

Top: *The waters of Rainy Lake, Ontario, are so calm that despite the fact that **DHC-3 Otter C-GMDG** (c/n 302) of Northern Wilderness Outfitters is moving, albeit slowly, through the water, it can still produce a perfect reflection. This aircraft was first flown in 1958 and served with the US Army. Operations included time in Vietnam from 1962 to 1971, with a period back in mainland USA for overhaul and maintenance. It was put up for disposal in 1973 and two years later saw it operating in Fairbanks, Alaska. In 1980 the aircraft was in Canada and since then it has flown fishermen to outpost cabins in the underpopulated north-west part of the vast Canadian province.*

Bottom: *Vermillion Bay, Ontario, is the location for this picture of **DHC-3 Otter C-FODV** (c/n 411) of Wilderness Air, which for thirty years has been flying hunters and fishermen to their lakeside lodges in remote wilderness areas during the summer months of May to October. The company advertises that it goes to areas few people have ever been and where the fish have never seen a lure. For those interested, the angler can find in the water such fish as walleye, northern pike, muskie, smallmouth bass and trout. As well as the getaway-from-it-all locations, stressed city dwellers may also be lucky enough to see the Northern Lights, or Aurora Borealis, and those who have seen the pulsating green sky at dusk will not forget it.*

Top: **DHC-3 Otter C-GPHD** (c/n 113) of Sabourine Lake Airways is landing with a cloud of spray on the lake of the same name. This Manitoba-based carrier is one of the many that operate summer-season flights to fishing lodges.

Bottomv: Anchorage's Lake Hood is the setting for **DHC-3 Otter N491K** (c/n 434) of Katmai Air. When seen out of the water next to a pick-up truck, the size of the Otter can be judged. The company name derives from Alaska's Katmai National Park, famous for its brown bears. Each year millions of salmon arrive from the Bering Sea to spawn, and they provide a feast for the large population of bears; there may be as many as fifty of these animals in as short a stretch as 2 miles of the Brooks River. Brooks Falls is the viewing point, and the company will fly tourists from Anchorage to a nearby lodge to stay overnight and watch the bears feed during the salmon season.

Below left: *Pictured in winter storage at La Ronge, Saskatchewan, is **DHC-3 Otter CF-PEM** (c/n 438), operated by Athabasca Fishing Lodge and owned by Cliff and Stella Blackmur since 1974. The company name derives from the lake of the same name, the largest in the Canadian province and regarded as one of the best 'big fish' waters in the world. Despite being in a remote area, some of the lodges have almost a four-star hotel rating; they have hot water, electric power, air-conditioning, maid service and full meals provided by on-site cooks with, of course, fresh fish on the menu.*

Below right: *Built in 1967, **DHC-3 Otter C-FVQD** (c/n 466) of Manitoba-based Sabourine Lake Airways is pictured with its Pratt & Whitney R-1340 turning as it moves across to dock at Sabourine*

Lake SPB. Capable of taking ten passengers, a number have been re-engined with Polish-built, Russian-designed ASh-621R radial pistons. This nine-cylinder powerplant produces up to 1,000hp and is the same unit that is fitted to the Antonov An-2 biplane.

Bottom: *Green Airways was set up in 1950 for the role of flying fishermen to lakeside lodges, and operates an all-piston-powered fleet from its Red Lake, Ontario, base. **DHC-3 Otter C-FMEL** (c/n 222) is pictured at the company base awaiting its next load of passengers.*

*Top: Pictured just seconds before touch-down at Vancouver's downtown seaplane base is **DHC-3 Turbo Otter C-GVNL** (c/n 105) of Harbour Air. This locally based carrier claims to be the world's largest all-seaplane airline. As well as its operations in Canada, it has an associate company on the Mediterranean island of Malta, whose aircraft was fitted with a wheeled undercarriage for the long ferry flight across the width of Canada, Greenland, Iceland, Scotland, England and France. The floats, having been shipped over, were refitted to the aircraft and flights started from Valetta to the nearby island of Gozo. The aircraft has been re-registered with the '9H' prefix of Malta.*

*Bottom: The much longer and sleeker-looking turbine power unit is apparent in this picture of Vancouver Island Air's **DHC-3 Turbo Otter C-GHAG** (c/n 214), seen at the company's Campbell River base. As well as the cheaper running costs, once the expense of the conversion is factored the turbine-powered Otter will fly at a 40mph greater cruising speed, climb goes up by 350 feet per minute, and both landing and take-off distances are reduced by a substantial margin, the Otter already having a STOL capability. As the PT-6A is much lighter than the R-1340, as much as 730lb (345kg) of extra revenue-generating payload may be carried.*

*Top right: As well as its expertise with the Beaver Kenmore Air of Seattle, Washington is equally at home operating the Otter. Unlike its DHC-2 fleet, all of its Otters are turbine-powered. Pictured on an international flight to Victoria, Vancouver Island, is **DHC-3 Turbo Otter N707KA** (c/n 106).*

Middle right: The same aircraft is pictured on a more recent visit to Kenmore Air Harbor while undergoing maintenance. It is apparent that the engine has been removed for servicing. One of the advantages of the turbine powerplant is that the time between overhauls is increased to 3,500 hours, which is a direct cost benefit to a company, as when an aircraft is down for scheduled servicing it is not earning any revenue.

*Bottom right: All the same problems regarding the cost of fuel and availability of AVGAS that affect the Beaver are relevant to the Otter. Again the answer is to re-engine with a turbine. However, the conversion involves just the powerplant – the airframe is not extended as was the case with its baby brother. The engine of choice is the Pratt & Whitney Canada PT-6A, and Vazar Aerospace, based at Bellingham, Washington State, produces a turbine conversion kit for the job. It includes such items as a new fuel system, a three-blade Hartzell propeller with full reverse capability, engine cowlings and various changes to the instruments. Pictured at Anchorage International Airport, Alaska, is **DHC-3 Turbo Otter N17689** (c/n 431) of King Salmon-based Katmai Express. This is one of more than eighty Vazar conversions.*

Top: *Pictured arriving at the Lake Union SPB in downtown Seattle is* **DHC-3 Turbo Otter N606KA** *(c/n 37) of Kenmore Air. From this central location the carrier operates to a total of fourteen destinations, including the international service to Vancouver Island.*

Bottom: **DHC-3 Turbo Otter N50KA** *(c/n 221) of Kenmore Air is painted as an advertising billboard for the NBC K5 Evening News Seattle television programme. It is becoming quite common for airlines to paint one of their fleet in special livery either to commemorate an anniversary such as fifty years of operations, a major sporting event such as a motor-racing Grand Prix, or to advertise a local attraction.*

Below: *When in January 1937 the first Beech 18 took to the air from Wichita, Kansas, nobody would have guessed that it would remain in production until the end of 1969. Its designed role was that of a feeder liner seating up to nine passengers. The powerplants were a pair of Pratt & Whitney R-985 nine-cylinder, air-cooled radial pistons. The start of the Second World War saw the aircraft in uniform and in mass-production, as it was ideal as a pilot trainer for men going from single-engined aircraft to four-engine bombers. One version was made with a glass nose to train bomb-aimers and navigators; this variant was known as the AT-11 Kansan. Following the war, many ex-military aircraft were converted for civil use and production resumed with more modern models. Pictured at Fort Francis SPB, Ontario, is **Beech 18 C-FZRI** (c/n 9940) of Rusty Myers Flying Services.*

Bottom: *The twin Beech has had many post-manufacture modifications from various aerospace companies, including the fitting of a tricycle undercarriage, a single fin to replace the distinctive twin tails, various new engines, including turboprops, and stretched fuselages. Companies doing the conversions have included American Turbine Aircraft, Avco, Dee Howard, Dumod, Hamilton Aviation, Airlines Training Inc, Pacific Airmotive, Rausch, Remmert-Werner, Volpar and, in France, SFERMA. Still in its normal configuration is **Beech 18 C-FRVL** (c/n 7835) of Rusty Myers Flying Services, pictured at its Fort Francis, Ontario, base.*

Top: *For more than sixty-five years Rusty Myers Flying Services has taken fishermen to the remote lakes of Ontario. The company is based at Fort Francis, a border town next to the US state of Minnesota. Its location means that, leaving cities such as Chicago, Milwaukee or Minneapolis in the morning, you could be out fishing by the end of the day. The company operates more than ten cabins, with Grayson Lake, at 240 miles, being the furthest, and Metionga Lake, at 155 miles, being the closest to base. Both of these locations also host moose-hunting trips as well as fishing. **Beech 18 C-FBGO** (c/n CA-215) is pictured tied up at the dock at base.*

Top right: *Pictured on a flight from Fort Francis, Ontario, is **Beech 18 C-FERM** (c/n CA-62) of locally based Rusty Myers Flying Services. Noticeable is the extra ventral fin under the tail, necessary to offset the increased side area due to the EDO floats, which have small rudders fitted to them; note also the rope used for docking as it trails in the slipstream. In the background many small lakes are visible in the huge open vista.*

Middle right: *Walsten Air Service, based in Kenora, Ontario, was founded in 1986 and offers amongst its services medical transfers, aerial photography, and sightseeing passenger and freight charters in both VFR (Visual Flight Rules) and IFR (Instrument Flight Rules) operations. Pictured at Kenora SPB is **Beech 18 C-FCUK** (c/n CA-181) with company titles on the rear fuselage.*

Bottom right: *Showalters Fly-In Service, based at Ear Falls, Ontario, is another of the companies operating fishing and hunting lodges in the Canadian wilderness areas of the vast province. **Beech 18 C-FZNG** (c/n CA-182) is pictured departing from Red Lake SPB.*

Below: *Pictured at its Nestor Falls, Ontario, SPB base is **Beech 18 C-FNKL** (c/n AF-378) of Northwest Flying. This carrier operates a seasonal (May to October) service to outback cabins, and as well as this nine-place aircraft it also operates a DHC-2 Beaver with six seats and a DHC-3 Otter with ten.*

Top left: *The morning light reflects on Red Lake Airways* **Beech 18 C-GEHX** *(c/n CA-112) as it taxies out from the dock across the flat calm lake to begin its take-off run. Following its first flight in 1952 the airframe saw service with the RCAF until 1970, when it was sold on to the civil register and over the following years had a number of owners and operators.*

Middle left: *Operating since 1985 from Campbell River SPB on Vancouver Island, British Columbia, Vancouver Island Air started operations with a single Cessna on floats and currently operates a total of three twin Beech aircraft. Pictured at its dockside moorings is* **Beech 18 C-FGNR** *(c/n CA-191).*

Bottom left: *Another picture of Vancouver Island Air's* **Beech 18 C-FGNR** *(c/n CA-191) with both of its Pratt & Whitney R-985 air-cooled radial pistons pounding as it departs from the company base at Campbell River for a charter flight down the coast to Nanaimo with a full load of nine passengers.*

Below: *During the early 1960s de Havilland Canada realised that it needed a replacement for the Otter. For the growing market of commuter operations, two engines would be a must, and the advent of the Pratt & Whitney Canada PT-6 turboprop was the powerplant needed for the new aircraft. This was to be the Twin Otter. The prototype first flew in May 1965, and the design was certified in both the USA and Canada the following year. The aircraft has retained the STOL characteristics of its predecessors and can be found on wheels, floats or skis. Pictured on approach to land at Victoria SPB, Vancouver Island, is* **DHC-6 Twin Otter 100 C-GQKN** *(c/n 94) of West Coast Air. It has a seating capacity of*

eighteen, and the company runs scheduled passenger services around the province of British Columbia, with Vancouver to Victoria being one of the most travelled.

Below right: *Pictured just about to touch down at the city-centre Vancouver SPB is West Coast Air* **DHC-6 Twin Otter 100 C-FGQE** *(c/n 40). It is a great boon to business travellers that they can fly into or out of a location in the heart of the city.*

Top: *The background buildings show how close the seaplane operators can bring passengers to the centre of a city. Pictured at Victoria, Vancouver Island, is **DHC-6 Twin Otter 200 C-GJAW** (c/n 176) of West Coast Air. The -200 series has additional space for luggage in the rear fuselage as well as in the nose, although the latter extension is not applicable to the floatplane, as the loading and unloading would be a problem in dock areas.*

Bottom: *Next came the -300 series, which had a large cargo door split into two parts, one having an integral air stair. In addition,* the fuel capacity was increased. The type was certified in April 1968, with deliveries starting the following year. More powerful PT-6 engines gave it a higher weight operation and increased the cruising speed to 182 knots (337kmph). Pictured on the move on Great Slave Lake, Yellowknife SPB, is **DHC-6 Twin Otter 300 C-GMAS** (c/n 438) of locally based Air Tindi. This company operates a mixed fleet of seven main types on both scheduled and on-demand charter operations in Canada's far north both from water, as pictured, and from land.*

Top: *First flown at the end of 1982, the Cessna Caravan has proved to be a great success in its utility roles in all parts of the world. It has seating for nine passengers and is powered by the very reliable and well-tried Pratt & Whitney PT-6A turboprop. As well as its fixed tricycle undercarriage, it will operate on skis or floats. Pictured in the water at its dock on Lake Hood, Anchorage, Alaska, is* **Cessna 208 1 N675HP** *(c/n 208-00289) of Rust's Flying Services. This carrier has a mixed fleet, including Beavers and Otters, and operates year-round for sightseeing flights, fly-in fishing, hunting trips and bear-viewing, all from the largest and busiest floatplane base right next to Anchorage International Airport.*

Bottom: **Cessna 208 Caravan 1 C-GATY** *(c/n 208-00305) of Yellowknife NWT-based Air Tindi is pictured powering along the waters of Great Slave Lake. The location is appropriate, as the name 'Tindi' means 'Big Lake', or 'Great Slave Lake' in the local Dogrib tongue. Founded in 1988, the company operates both freight and passenger services in a mix of aircraft from three to forty-six seats. When Cessna designed the Caravan it fitted a large cargo door, as so many flights made by the company are a mix of freight and passengers. During the long cold winters, when the lake is frozen, the aircraft will operate on skis replacing the floats until the next spring. Flights are conducted from both the main airport as well as the SPB on the lake.*

Top left: *Along with the Beaver, still flying to this day, is another Canadian aircraft that predates the de Havilland design by more than a decade. Bob Noorduyn started work on the aircraft towards the end of 1934, and asked potential customers and their pilots what they would like to see. From this consultation came a convention high-wing, single-engine aircraft, the Norseman. It had a wide cabin and a side loading door able to take a fuel drum. One pilot was joined by up to nine passengers or their weight in cargo. One other important factor was that the wing was supported by struts, which had an additional use in water operations: they made docking easier as the aircraft could be manhandled by the ground staff. Pictured on its take-off run at Sabourine Lake, Manitoba, is* **Noorduyn Norseman VI C-FJEC** *(c/n 469) of Sabourine Lake Airways.*

Middle left: *The same aircraft is pictured making a flypast over the lake as it carries a family and all their food supplies to the remote area where they live.*

Bottom left: *The first flight of the Norseman was in November 1935 from Montreal's St Lawrence River, in float configuration. The powerplant was a 420hp Wright Whirlwind radial air-cooled piston engine. The first production batch was the Mark II, with just a small increase in the engine's power output. The Mark III sported a Pratt & Whitney 420hp Wasp, and again the type was thought to be underpowered. The Mark IV Norseman had a 550hp P&W R-1340*

Wasp engine, and all subsequent variants used this powerplant as it proved to be the ideal one. This mark first flew in November 1936, a year from the type's first flight to get the right engine and airframe mix. Pictured taxiing into base at Lac Seul, Manitoba, is **Norseman VI C-FKAO** *(c/n 636), then operated by Gawley's Little Beaver Lodge, a seasonal operator to remote fishing cabins.*

Below: *The start of the Second World War saw the Norseman join the military with service in both the RCAF and in the USA, the latter operating approximately 750 aircraft with the designation C-64, having a higher all-up weight and increased fuel capacity. This variant became the Norseman Mark VI; the manufacturer held back using the designation 'Mark V' as it wanted to save this for the end of hostilities, the 'V' signifying Victory, based upon wartime Prime Minister Winston Churchill's 'V-sign'. Pictured at its dock at Red Lake is Green Airways* **Norseman Mark VI C-FOBE** *(c/n 480). The company has an all-piston-powered fleet, including Beavers and Otters.*

Bottom: *Pictured out of the water at Yellowknife Airport, NWT, is* **Norseman Mark V CF-SAN** *(c/n N29-29) operated by Buffalo Airways. Although certified for commercial operations, the airline's owner, Joe McByran, uses it mostly as his own personal floatplane. This pilot has found fame as the boss of Buffalo Airways in the TV documentary Ice Pilots.*

Top: *It is almost rare to see a Norseman on wheels rather than floats. Pictured at Oshkosh, Wisconsin, is ex-US military* **Noorduyn UC-64 Norseman CF-LZO** *(c/n 535). One of the great mysteries of aviation, which will probably never be solved, concerns the Norseman. On 15 December 1944 Major Glenn Miller, the famous bandleader, climbed aboard a Norseman at RAF Twinwood Farm near Bedford bound for Paris to make arrangements to bring his musicians over for a series of Christmas concerts for the troops in France. The aircraft was never seen again, nor was wreckage ever found. Many theories have been put forward, from the logical to the ridiculous, but no confirmed answer has ever been found.*

Bottom: *For those who want a bush plane without the cost of a newly refurbished Beaver or equivalent, there is the Murphy Moose. The only snag is that the manufacturer does not build it, but provides it in kit form for home builders to tackle the job themselves. The powerplant for this aircraft, which has a performance not dissimilar to the Beaver, is a Russian-built Vedeneyev M-14P nine-cylinder air-cooled radial, which with supercharging produces 360hp. Also available is a six-cylinder in-line Lycoming with a power output of 260hp. The aircraft can seat up to six people. Pictured on Lake Winnebago, Wisconsin, is* **Murphy Moose N622D** *(c/n 130). The performance figures for the type are a maximum speed of 175mph (282kmph), a cruise speed of 155mph (249kmph), a range of 600 miles (965km), and a service ceiling of 15,000 feet (4,575m).*

Top: *The first of a series of four amphibian aircraft from Grumman, based at Bethpage, New York, flew in May 1937. The Goose, as it was known, was powered by a pair of Pratt & Whitney Wasp Junior air-cooled radial pistons with an output of 450hp each. Six passengers with a crew of two could be transported. During the Second World War 345 airframes were built for the US Navy, the RAF and the RCAF, as well as for the US Coast Guard. Pictured at Red Lake Airport, Ontario, is* **Grumman G-21 Goose C-GRCZ** *(c/n B-49), owned by a resort hotel to ferry guests from the airport.*

Bottom: *With a large number of surplus airframes following the war's end, a number of companies converted them to civil operation. McKinnon Enterprises made the most radical variants, including one with four engines. Pictured at Oshkosh, Wisconsin, is* **McKinnon G-21G Goose N70AL** *(c/n 1226), which has been re-engined with a pair of 680shp P&W PT-6A-27 turboprops. As can be seen when compared to the original engines of the Goose, the new shape is quite distinctive.*

Top: *The second of the four Grumman amphibians was the Widgeon. This had the same layout as the Goose, but was smaller with room for a pilot and four passengers. Power came from a pair of Ranger in-line engines of 200hp each. It first flew in June 1940 at the manufacturer's Bethpage site, and saw wartime service with the USN, the USCG and the Royal Navy, in whose use it was called the Gosling. Pictured at Oshkosh, Wisconsin, is* **Grumman G-44 Widgeon N68102** *(c/n 1351), a privately owned example.*

Bottom: *As well as post-war US production, a line to manufacture the Widgeon was set up in France by SCAN (Société de Construction Aero Navale) at La Rochelle. Pictured at Oshkosh, Wisconsin, is* **SCAN 30 (G-44A Widgeon) N540GW** *(c/n 3), one of the French-built examples that has found its way back across the Atlantic and is now privately owned. Just forty aircraft were built in France; production ran into problems with the supply of the Ranger engine, to such an extent that the first French prototype had de Havilland Gipsy Queens on its first flight. The SCAN 30s still flying can be found with a number of different powerplants.*

Above: *The last of the four Grumman amphibians, the Albatross, is by far the largest. It first flew from the Bethpage plant in October 1947, and was soon in service with the USAF and the US Coast Guard in an air-sea rescue role and based in world-wide locations. As well as the American military, the Albatross served with the armed forces of Brazil, Argentina, Greece, Canada, Chile, Germany, Indonesia, Italy, Japan, Malaysia, Mexico, Norway, Pakistan, Peru, the Philippines, Portugal, Spain, Taiwan, Thailand and Venezuela, and many were from ex-US service stocks. Pictured under tow at Oshkosh, Wisconsin, is* **Grumman HU-16E Albatross N29853** *(c/n G-335), owned by the energy drink company Red Bull, based in Salzburg, Austria, which sponsors a number of aviation assets.*

Below: *The Albatross was powered by a pair of Wright R-1820 nine-cylinder air-cooled radial piston engines with a power output of 1,425hp. The ability to alight on either land or water, together with its large size, has seen some privately owned airframes converted into luxury corporate-interior-styled flying yachts. Pictured with both engines turning,* **Grumman HU-16B Albatross N98TB** *(c/n G-243) taxies to its parking position upon arrival at Oshkosh, Wisconsin.*

Top: *It is the passion of some owners of classic ex-military aircraft to retain either the original scheme in which it operated or to put it in a livery that shows its former heritage but may not be totally accurate.* **Grumman HU-16C N7025N** *(c/n G-409) wears US Navy titles and the serial 141262 as it retracts its undercarriage while taking off from runway 36 at Oshkosh, Wisconsin. The aircraft is privately owned and Michigan-based.*

Bottom: *Pictured at Lake Union Sea Plane Base, Seattle, are a pair of Kenmore Air aircraft, providing a chance to compare the size of the two de Havilland Canada classics, the DHC-2 Beaver and its big brother, the DHC-3 Otter.*

Top: *In this typical view of Campbell River Sea Plane Base are three of the types operated by Vancouver Island Air, from left to right the DHC-3 Otter, Beech 18 and DHC-2 Beaver. Although operated by the same carrier, each has quite a different livery style.*

Bottom: *The last picture in this section shows how waiting for a flight can be, but sadly is not for most people. It shows passengers waiting to board their Green Airways floatplanes at Red Lake Sea* Plane Base, Ontario. *How wonderful to be out in the open air on a fine day in a picturesque part of the world, and to board, without hassle, a classic aircraft and fly off water to spend the next week in splendid isolation. Not for these passengers the long check-in queues, the baggage searches, the x-ray machines, and removal of shoes, belts and other items of clothing – truly an airport experience with a difference!*

3 Preserved aircraft and pleasure flying

The concept of preservation and museums has been with the human race for centuries. It is therefore right and proper that aeroplanes that have changed the way we live and have shrunk the world should be preserved for future generations to see. There can be no better way of doing this than keeping the aircraft in airworthy condition and flying them on a regular basis to let the public of today fly in them and see what it was like in the often-called 'golden years' of aviation. That was a time, however, when only the rich and influential could afford the fares and time to travel the world.

To the general public the thought of going for a flight in an aeroplane up to seventy years old might have them running for the exit, but enthusiasts and anyone with a spirit of adventure will leap at the chance. The smile upon their faces when they disembark says it all.

The preservation of airliners is fortunately a growing trend, and the range varies from the evergreen classic Douglas DC-3, perhaps the most common offering flights, to a genuine German-built Second World War vintage Junkers Ju 52/3m tri-motor operated by Swiss-based Ju Air, without a doubt the best way to see the Alps.

Some of the major airlines of the world take their history and heritage seriously, none more so than German flag-carrier Lufthansa. As well as the Ju 52/3m that it operates for promotional and corporate events, it has bought three examples of the ultimate variant of the beautiful Lockheed Constellation, the L-1649 Starliner, and is currently working to put one back in the air in the next few years. Together with the work on the airframe, being done in the US state of Maine, Lufthansa is also training pilots to fly the aircraft. Today's airline pilots will find the differences between their normal Airbus or Boeing types and a classic propliner so great that they will need a lot of conversion work to be able to operate it safely.

As well as the thrill of a flight, many operators add to this by taking the aircraft to an event such as an air show or to view some of the wonders of the world. Places like the Grand Canyon in Arizona cannot be as fully appreciated from the ground as they can from the air. The tallest waterfall in the world, Angel Falls in Venezuela, is in a very remote part of that vast country, and again the best way to see it is from the air.

To sit in the passenger seat of a true classic propliner as the engines burst into life with clouds of smoke and flames from the exhaust and vibration throughout the whole aeroplane is something that everyone should experience at least once in their life.

Top: *The Lockheed Constellation is for many aircraft enthusiasts the most elegant airliner of all time. Its first flight was in January 1943 from the Lockheed Air Terminal, Burbank, California, but the pilot in command was Boeing's Chief Test Pilot, Eddy Allan. The reason for this was that he had far more multi-engine experience than his opposite number at Lockheed, and was therefore on loan for the occasion. The Constellation went through four different models, the L-049, the L-749, the L-1049 (now called Super Constellation), and finally the L-1649 (named Starliner). Pictured in the sky over Payerne, Switzerland, is* **Lockheed L-1049 Super Constellation N73544** *(since re-registered HB-RSC) (c/n 4175). It is owned by Swiss-based Super*

Constellation Flyers Association and carries the name of Swiss watchmaker Breitling, one of the main sponsors. This aircraft was built for the USAF in 1955 and served with MATS (Military Air Transport Service) and later with both the Mississippi and West Virginia Air National Guard before being stored in 1972. Its first civil role found it fitted with spray bars to combat budworms in spruce trees. Following this several schemes were proposed but failed to materialise, and the aircraft's condition deteriorated with no flights between 1984 and 1994, during which time a long restoration took place. It made the journey to Switzerland ten years later.

Below: *Pictured taxiing down the runway at Avalon, Victoria, is* **Lockheed L-1049F Super Constellation VH-EAG** *(c/n 4176), owned and operated by the Historic Aircraft Restoration Society (HARS) based at Illawarra, Australia. Built in 1955, it served with the USAF and later the Pennsylvania ANG. It was retired and stored at Davis-Monthan AFB, Tucson, Arizona, in 1977, and in the following years several plans for its future fell by the wayside. In 1992, with its ownership transferred to the Australians, it moved next door to the Pima County Air Museum for the start of its restoration programme. This culminated two years later when it took to the air for the first time in seventeen years. More work continued, including the fitting of tip-tanks, and at the start of 1996 it made a twelve-day trans-Pacific flight to its new Australian home.*

Top: *Forever known as the 'Connie', the L-1049 is powered by four Wright Cyclone eighteen-cylinder R-3350 air-cooled radial pistons with an output of 2,700hp each. Pictured running all four at Avalon, Australia, is the HARS* **Lockheed Super Constellation VH-EAG** *(c/n 4176).*

Bottom: *As well as the thrill of flying in propeller-driven aircraft, one of the other advantages is that they are ideal for sightseeing from the air. The DHC-6 Twin Otter turboprop (as covered in Section 2) has many uses, and one specialised variant is the Vistaliner. This version has extra-*

large windows giving all nineteen passengers, sitting in a 2-1 layout, an excellent panoramic view; in addition, a four-blade noise-suppressing propeller is fitted, giving a far quieter ride and reducing the noise footprint of the aircraft. This last item is very useful for Grand Canyon Airlines, which offers passenger flights over one of the natural world's most spectacular sites. Pictured at Monument Valley Airport, Utah, is **DHC-6 Twin Otter 300 Vistaliner N173CG** *(c/n 295). As can be seen, the windows are far larger than the standard airframe. Conversions to Vistaliner standard are by Las Vegas-based Twin Otter International.*

Top: *Last of the line of the Douglas passenger propliners, the DC-7 first flew in May 1953 and entered operations with American Airlines later that year. The type's 'front-line' service with the flag-carrying airlines of the world was short-lived, as within a few years the early jets had arrived in the form of the Boeing 707 and, from Douglas, the DC-8. The airline customer wanted jets on the main routes and got them, while the DC-7 found itself operating for smaller airlines or as a freighter. One other use for the older airframe was the travel club. People have always wanted to travel and the cost was often far too high for many. Out of this grew clubs where the members owned an airliner, which could be used to transport them to far-flung destinations. One such, founded in 1972 at St Paul, Minnesota, was the Twentieth Century Travel Club, which purchased* **Douglas DC-7B N836D** *(c/n 45345).*

However, operations never got under way and the aircraft remained parked up with just the occasional engine run. It is pictured at St Paul – but that wasn't the end of the story…

Bottom: *At the end of 2003 the owner of Florida Air Transport purchased the DC-7 after inspecting it at St Paul. August of the following year saw the aircraft take to the air again for the first time in more than thirty years following some months of restoration for a ferry flight. It was then flown to Opa Locka, Florida, for the start of its long-term restoration and repaint in the livery of its first owner, Eastern Airlines. Six years later, in July 2010, the aircraft was ready, and* **Douglas DC-7B N836D** *(c/n 45345) is pictured at Oshkosh, Wisconsin, ready to take passengers once again.*

Top: *Like the last of the great piston-powered airliners, the first of the western-built turboprops also had a relatively short life in the front line with major airlines. It was the paying public's yearning for jets that saw them off. The Lockheed Electra first took to the sky from Burbank, California, in December 1957, and entered Eastern Airlines service in January 1959. Confidence in the type was lost following a number of accidents that were found to have been caused by excess vibration leading to a failure of the wing root. Lockheed cured the problem, but the march of jets saw the Electra moved from the prime routes. The last passenger-carrying Electra operator was Reeve Aleutian Airways, based in Anchorage, Alaska. Most aircraft were operated in a combi-mode, with part of the fuselage configured with passenger seats and the rest cargo. Pictured landing at base in late evening sun is* **Lockheed L-188C Electra N1968R** *(c/n 2007). The company recognised the desire for flights in the type, so operated Electra Fan Flights, which* allowed for flat-rate prices to any location in the company network on a space-availability deal as long as no full-price passenger wanted the seat.

Bottom: *CHC Helicopters, based in Vancouver, Canada, is the world's largest company specialising in to-and-from oil and gas rig transportation, civilian air-sea rescue, and the service and repair of rotor-winged craft. The fleet of more than 250 aircraft is to be found operating in more than thirty nations. One of the company's fixed-wing aircraft,* **Convair 580 ZS-LYL** *(c/n 39), is pictured at Cape Town, South Africa. Its Allison 501 turboprops are turning as it gets ready to depart with a party of British aviation enthusiasts for a sightseeing flight around the area. This airframe was built in 1952 as a model 340 for Delta Air Lines with piston power, and was converted to a turbine-powered aircraft in 1969.*

These three pictures, taken at Berlin-Schönefeld, show **Douglas DC-6B N996DM** (c/n 45563) in the livery of Salzburg-based energy drink company Red Bull. Red Bull has a strong aviation presence, including a museum with flying warbirds, the Red Bull Air Racing season, and fixed-wing aircraft and helicopters based in the USA for air show work. The pride of the fleet has to be the big four-engine Douglas. Delivered in 1958, it was the last but one DC-6 built. It served first the national airline of Yugoslavia, JAT, followed by that nation's air force, which donated it to the Zambian Air Force in 1975, and it stayed in Africa when it moved to Namibia in 1994. The year 2000 saw its sale to Red Bull and its move to Europe. There it received extensive maintenance and restoration, making the airframe the smartest-looking DC-6 in the world today. And to match the exterior it has a VIP interior installed.

Top left: *Bathurst Vintage Joy Flights is based at Bathurst, New South Wales, Australia, and operates one of the few Australian-designed, Australian-built aircraft, the de Havilland Australia Drover, a basic utility transport seating up to eight passengers. Power came from three Gipsy Major piston engines of 145hp each. The undercarriage was fixed and had a tail wheel. It first flew in January 1948 and its most famous operator was the Royal Flying Doctor Service. The aircraft was configured for a pilot, two medical personnel and two stretcher positions. Pictured at Avalon, Victoria, is **DHA-3 Drover VH-ADN** (c/n 5009). Just twenty airframes were constructed.*

Middle left: *Developed from the earlier model 2-0-2, the Martin 4-0-4 was a post-Second World War airliner that the manufacturer hoped would equip the airlines of the world in the expected post-war passenger boom. However, the type was in direct competition with the Convair 240, and both had to face the huge numbers of surplus Douglas DC-3s being released by the military. The 4-0-4's first flight was in October 1950 from the company's Baltimore site. Service entry was on 15 January 1952 with an official inauguration by both TWA and Eastern Airlines, although both carriers had operated the type on ad hoc services during the two previous months. Pictured at Oxnard, California, is **Martin 4-0-4 N636X** (c/n 14135) in the period livery of Pacific Air Lines. It has a sixteen-seat executive interior and is owned by Jeff Whitsell, operating as Airlines of America. It is being prepared for a tour of the western USA by British aviation enthusiasts.*

Bottom left: *Prestwick-based Scottish Aviation has during the post-war period designed and built two different aircraft types. The first was the single-engine Pioneer, a utility and communications aircraft with an STOL performance. Following this, in 1955, came the Twin Pioneer, with the same role and similar performance in short-field operations. The RAF operated the aircraft as a light transport and the fleet spent almost all of its life east of Suez. Powerplants for the Twin Pioneer were a pair of Alvis Leonides air-cooled radial pistons with an output of 640hp in the Mark 3. Pictured at its Coventry base is **Scottish Aviation Twin Pioneer 3 G-APRS** (c/n 561), one of Atlantic Airlines' historic fleet and used for pleasure flying. The colour scheme it wears is the 'Raspberry Ripple' of the Royal Aircraft Establishment, and it also served on the strength of the Empire Test Pilots School. It is quite common for such schools to have what could be deemed an odd choice of aircraft, but few military pilots would have any experience in flying a STOL twin-piston-powered tail-dragger, so it is an excellent teaching aid for their course.*

Below: *Operating pleasure flights in Australia was a carrier with the right name for the type of aircraft flown. Pictured at Coolangatta, Queensland, is Twin Pionair Airlines' **Scottish Aviation Twin Pioneer VH-AIS** (c/n 540). It is at base awaiting a passenger charter flight around the local area on an untypically damp day.*

Top left: *South African travel company Rovos, based at Pretoria-Wonderboom, operates both trains and aeroplanes, offering high-quality tours including wildlife safaris and journeys to some of the wonders of the African continent. Pictured arriving at Livingstone, Zambia, is **Douglas DC-4 ZS-AUA** (c/n 42934). It has shut down its outer two Pratt & Whitney R-2000 air-cooled radial piston engines as it taxis to its parking stand to unload its passengers, who have arrived to view the nearby Victoria Falls, one of the natural wonders of the world. As well as the Rovos Air title, the aircraft carries the name 'Flying Dutchman', a tribute to the previous operator, the Dutch Dakota Association.*

Middle left: *Pictured following the DC-4 into Livingstone, Zambia, with more passengers to see Victoria Falls, is **Convair 440 ZS-BRV** (c/n 215), also operated by Rovos Air. This Convair has a configuration for forty-four passenger seats and is still piston-powered with original Pratt & Whitney R-2800 radials. It was built in 1954 and served with the USAF as a C-131D, then in 1987 was retired to be stored at Davis-Monthan AFB in Tucson, Arizona. It emerged in 1992, joining the US civil register, and was sold later that year to an airline in Bolivia.*

Bottom left: *The same aircraft is pictured in its Bolivian identity; **Convair 440 CP-2236** (c/n 215) is at its Cochabamba base in the livery of Lineas Aereas Canedo, a company that was established in 1978 and has only operated piston-powered aircraft since that time. This Convair was one of a pair purchased in 1992 and had a seating capacity of forty passengers in a two-class arrangement. The two aircraft operated all over Bolivia and were leased to the national flag-carrier Lloyd Aero Boliviano for a period of time. In July 2001 they were both sold to Rovos Air and ferried on the long journey from Bolivia to South Africa. The route was Cochabamba to Porto Suarez, then to Natal on the coast of Brazil. Ferry tanks were fitted for the 10hr 40min South Atlantic crossing to Abidjan, the capital of Ivory Coast, then on to Namibia and finally its new home in South Africa.*

Below: *When in June 1948 the Russian forces blocked road and rail access to Berlin in a plot to gain control over the divided city, they did not expect the western allies to reply by flying in all the supplies needed for a whole city. Over 200,000 flights took place bringing every item needed for the civil population. The blockade ended in May the following year and by then more shipments were arriving by air than had previously been transported by ground. To commemorate this event the Berlin Airlift Historic Foundation was set up. They fly to air shows in the USA to educate the public and to raise funds to keep their aircraft in the air. Pictured at Detroit – Willow Run is **Douglas C-54 Skymaster N500EJ** (c/n 27370) operated by the group. First flown in 1945 this aircraft served with the US Navy and the Marine Corps before being stored in 1973. It later spent a period operating with a Toronto-based Canadian freight company before joining the foundation in 1994.*

Above: *One of the projects of the Berlin Airlift Historic Foundation raise funds for is to bring their C-97 back into the air. Pictured at New York's Floyd Bennett Field is* **Boeing C-97 N117GA** *(c/n 16749) in the markings of YC-97A 45-59595 the only one of its type to operated in the Berlin Airlift. Its real ex-military identity is 52-2718. Power for this aircraft comes from four Pratt & Whitney R4360 radial pistons with a power output of 3500hp each. The group acquired the airframe in 1996 and following work at a number of sites is planning to join the C-54 on the air show circuit around 2013. This aircraft started its life in 1954 with the USAF's Strategic Air Command as a KC-97G flying tanker to refuel their bombers.*

Below: *One of the most iconic German aircraft of the Second World War was the Junkers 52 transport. It flew for the first time in October 1930 with a single engine; the design was subsequently altered to take three motors, and this variant took to the air in April 1932. The powerplants depended upon the operator, and at least*

six different ones were fitted. The aircraft flew with airlines from Peru to Poland and Austria to Australia. The German flag-carrier Lufthansa has a strong interest in its history and in 1984 acquired its own Ju 52 to restore and operate for promotional purposes. Pictured at Berlin-Schönefeld is **Junkers Ju 52/3m8e D-CDLH** *(c/n 130714), carrying the airframe's original registration of D-AQUI, when it was operated on floats in Norway during 1943. Following the conflict it stayed in that country and operated as a passenger airliner until it was sold in 1957 to Ecuador; it operated there until 1963, when it fell into disrepair. Rescue came in the form of American aviation author Martin Caidin, who flew it in Florida under the name 'Iron Annie' until its acquisition by Lufthansa.*

Top: *Once owned by the Confederate Air Force is* **Casa 352L N352JU** *(c/n 67). This aircraft is a licensed Spanish-built version of the famous Junkers Ju-52/3m tri-motor transport of World War II. It is currently owned by the Virginia-based Fighter Factory. It is one of 170 built by Casa (Construcciones Aeronauticas SA) and the type remained in service with the Spanish Air Force into the 1970s when a number were sold to collectors and museums who almost without exception painted them in German war time markings. This aircraft is pictured at Virginia Beach Airport whilst crew training.*

Bottom: *Operated by Air Heritage of Beaver Falls, Pennsylvania is* **Fairchild C-123K N22968** *(c/n 20113) in its former USAF markings of 54-0664 when pictured at Andrews AFB, near Washington DC. The C-123 became infamous during the Vietnam War when it was used in the 'Ranch Hand' operations. The purpose of these was to spray a chemical defoliant over the Vietnamese jungles to deprive the Viet Cong of their natural cover. The most used chemical was code named Agent Orange and proved to be very toxic in the concentrations used and has caused many deaths and subsequent birth defects.*

Top: *The Swiss Air Force bought three Ju 52s at the end of 1939 and flew them until 1981. Their roles were to train observers and radio operators, but transport and parachute work was also carried out. Rather than scrap the aircraft or put them in a museum, a novel solution was found to raise money for the Swiss Air Force Museum at Dübendorf near Zurich. The three aircraft would be operated by newly created Ju Air and offer pleasure flights through – not over – the Alps. As might be expected, all three were in excellent condition and are still powered by original BMW 132A3 radial piston engines.* **Junkers Ju 52/3m (G4E) HB-HOP** *(c/n 6610) is pictured flying at its Dübendorf base and shows off well the corrugated load-sustaining metal skin. The aircraft are fitted with seventeen seats and still have the cord along the cabin roof to which parachutists clipped their static lines. They have also appeared in a number of Second World War films painted in Luftwaffe markings.*

Bottom: *Of all the classic old airliners that are available to fly in, the venerable DC-3 is by far the most common and accessible. It is still used to this day for regular scheduled passenger flights in both the northern and southern hemispheres. In the African nation of Tanzania Indigo Aviation has recently started services from Dar Es Salaam to a number of locations in the country. The most famous operator is Buffalo Airways, based in Yellowknife, Northwest Territories. As well as charter trips to all parts of the north, the company runs a daily schedule to and from Hay River across the Great Slave Lake in a journey time of 50 minutes. Pictured at Hay River is* **Douglas C-47 C-GPNR** *(c/n 13333) with its starboard engine running, which was requested by the passengers so that they could see their aircraft with the propellers turning before they boarded. Buffalo is after all a very friendly and obliging airline.*

Top Right: *Amongst that most common aircraft is to be found one of the very rare variants, this being one of the very few Super DC-3s. The version has already been covered (see Section 1), and most of the tall-tailed airframes to be found are ex-military C-117s. However, Trans Northern, based in Anchorage, Alaska, operates one of the only surviving civil-built aircraft, which, due to the volume of surplus C-47s at the end of the war, failed to find any sales except for three to Capital Airlines. Pictured on the ramp at Kenai, Alaska, is* **Douglas Super DC-3S N30TN** *(c/n 43159); it is fitted out with nineteen seats in a 2-1 layout and is on a passenger pleasure flight charter from base.*

Middle: *Despite the company name, Shortstop Jet Charters offers passenger charters from its Melbourne-Essendon base in the classic DC-3 as well as more modern types such as a Learjet or a Citation. Photographed on the ramp at base is* **Douglas C-47 VH-OVM** *(c/n 33102); this airframe was built in April 1945 and first served with the Royal Australian Air Force. It took civil markings in 1989 and has been operated by Shortstop ever since. Among the services the company offers are breakfast, lunch or dinner flights, pleasure trips around the city of Melbourne, and food and wine day trips to Wangaratta to visit the Brown Brothers Epicurean Centre.*

Bottom: *The title Air Nostalgia is perhaps the best name for a DC-3 operator. Pictured on the ramp at the company's Melbourne-Essendon base is* **Douglas C-47 VH-TMQ** *(c/n 32884), owned by Australasian Jet. Like Shortstop, this company also offers charters, and its list includes the Coonawarra wine region, golf trips to remote courses such as King Island on the Twelve Apostles coastal formation, or the best for an aviation enthusiast, a flight to the Temora Aviation Museum in New South Wales. This airframe was another built in 1945, serving first with the RAAF, then a number of Australian owners and operators.*

Top: *This photograph on the ramp at Melbourne-Essendon shows aircraft of both the Australian DC-3 pleasure flight operators, Air Nostalgia and Shortstop Jet Charters.*

Bottom: *The DC-Association is a group based in Helsinki, Finland, whose aim is to maintain and fly an historic DC-3 for the benefit of members and to tour the aircraft to both Finnish and European air shows. Pictured on the ramp at base is **Douglas DC-3 OH-LCH** (c/n 6346) in the period livery of Finnish Airlines, which operated it from 1948 to 1960. This aircraft was originally for Pan American when it was built in 1942, but with the war ongoing it joined the US military until 1948. Following its Finnish Airlines service it was stored, had a cargo door fitted, then began a second career in uniform with the Finnish Air Force, which operated it until 1984, when it was re-registered with its original identity.*

Top right: *Flygande Veteraner (Flying Veterans) is an organisation based in Stockholm-Bromma, Sweden, and operates and maintains a vintage DC-3 that has spent most of its life in Scandinavia. Pictured visiting an air show at Tampere, Finland, is **Douglas C-47 SE-CFP** (c/n 13883), in the period livery of SAS (Scandinavian Airlines System) that it once wore. The aircraft was built in 1943 and served with the US military until it joined SAS in 1946. Between 1960 and*

1984 it was back in uniform with the Swedish Air Force before the current owners took it over. It has a summer schedule of up to ten flights per month for the group's members; these are mostly local sightseeing, but trips to other European airports are featured.

Middle right: *Coventry-based Air Atlantique has operated a number of DC-3s for pleasure flying. Pictured at base is **Douglas C-47 Dakota G-AMPZ** (c/n 32872), wearing the colours of RAF Transport Command and the serial KN442, its original RAF serial when it was delivered to that unit in 1945. It was sold to civil operators in 1952 and served with many different airlines over the years before joining the present one in 1990. The RAF markings are to commemorate the 50th Anniversary of the Berlin Airlift in 1998.*

Bottom right: *In Holland the Dutch Dakota Association was set up to preserve two DC-3s and offer flights, either domestic sightseeing from its Lelystad base or trips to European events or cities. Pictured on a flight to RAF Cottesmore, for an air show, is **Douglas C-47 PH-DDZ** (c/n 19754) in the period colours of Dutch airline Martins Air Charter. This company, a subsidiary of KLM, now operates an all-jet fleet under the name Martinair. This 1944-built aircraft served with the US military until 1961, then with a variety of operators in a number of countries until it joined the DDA in 1987.*

4

First-generation turboprops still in operation

It has to be admitted that the turbine engine is a more efficient powerplant than the big radial. It uses less fuel, that fuel is cheaper and in common supply, it requires less maintenance and is more reliable; finally, it is lighter than pistons of equivalent power output, which translates into more revenue-generating weight that the aircraft can carry, be it passengers or cargo. It was therefore no wonder that as soon as they were able, the aircraft manufactures were hanging turbines from their designs.

Many of the first-generation turboprop aircraft were on the quest to produce the so-called 'DC-3 Replacement', although as has been proven over the years the only real DC-3 replacement has been another DC-3. In the UK the Avro (later Hawker Siddeley) 748 and the Handley Page HPR-7 Dart Herald competed in this market. The Avro design was by far the more successful in sales terms, and can still be found around the globe earning revenue for its owners. Its follow up, the BAe ATP, is currently finding favour in the cargo market, with fewer and fewer configured for passengers.

In Holland the Fokker company was the first to have its aircraft, the F.27 Friendship, in the air and in service, and both original Dutch-built and licensed American versions can still be found earning their keep, while the second-generation Fokker 50 has taken the design on to new sophistication.

Japan is not well known for aircraft production; most of the types built have been for its own air force and not sold to any other country. The one important exception to this was the YS-11, which did sell world-wide, although few are in commercial service today.

Russia, while not needing a DC-3 replacement – although it did have a licensed variant, the Li-2 – was more concerned with replacing the early Ilyushin pistons, the IL-12 and 14. From the Kiev-based Antonov company came a range of early turbine-powered aircraft, starting with the An-24.

When it came to the larger airlines Russia was able to operate its turboprops in front-line service on the prime routes for far longer than the west, as 'customer choice' was not a phrase in the lexicon of Aeroflot. The IL-18 airliner was able to fly the prime routes while in the west the Lockheed Electra and the Bristol Britannia were pushed from many routes by the demands of the paying passengers, who wanted pure jets.

Almost all of the early turbine-powered aircraft can still be found – just listen at your local airport for the high-pitched whine.

Top: *Like the mythical quest to turn base metal into gold, finding an aircraft to replace the DC-3 has foxed many great aircraft manufactures. Of the ones built, some have been successful in their role, but the DC-3 has always proved that the only true DC-3 replacement is another DC-3. In terms of numbers sold, the Dutch company Fokker has produced the most commercial success with its design. First flying from its Schiphol, Amsterdam, factory in November 1955, it entered service with the Irish flag-carrier Aer Lingus at the end of 1958. Power for the new aircraft was a pair of Rolls-Royce Dart turboprops. Pictured at Miramar, San Diego,*

California, is **Fokker F.27-400M Friendship 85-01607** *(c/n 10652) in the markings of the Golden Knights, the US Army parachute display team. The type is known in the US military as the C-31A, with just two purchased in 1985 for this particular role.*

Bottom: *With a production run of 580-plus aircraft built by Fokker, and more than 200 built under licence in the USA by Fairchild, the F.27 has been in service all over the world and can still be found in many roles. As passenger airlines have moved to newer equipment, many airframes have been converted for cargo-carrying. The high wing of the design means that the loading height to the fuselage is quite low, making the process simple. Pictured at Vancouver, British Columbia, is **Fokker F.27-600 Friendship N729FE** (c/n 10385) in the colours of the giant US small parcel carrier Federal Express. Medium-sized aircraft such as this are used to feed parcels from towns not big enough to have a direct feed to Memphis, the company's main hub; they arrive at cities such as Vancouver and have their cargos transferred to large jets for transport to the hub.*

Top: *The most successful British-built DC-3 replacement has been the Avro (later Hawker Siddeley) 748. Like many of the same class it was powered by a pair of Rolls-Royce Dart turboprops with a power output of 1,740shp (the Series 2 airframe Darts produced 2,105shp). The aircraft was first flown from Woodford, Cheshire, in June 1960, and the last of just under 300 UK-produced examples was delivered in 1989. As with the F.27, the 748 was produced under licence, not in the US but in India by Hindustan. Pictured at Fairbanks, Alaska, is* **Avro (HS) 748 -276 Series 2A C-FAGI** *(c/n 1699), of Air North Airlines, based in Whitehorse, Yukon, which is configured for a passenger load of forty people. These aircraft are used for scheduled services between the company base to Dawson City, Old Crow, Inuvik and, on a seasonal basis, for an international route to Fairbanks.*

Bottom: *The first commercial user of the Avro (HS) 748 was British independent airline Skyways, which commenced revenue services in December 1961. As well as the airlines of the world the 748 has found favour with air forces as both a standard transport and also in a VIP role. The Queen's Flight of the RAF used the type for some years, flying both members of the Royal Family and government ministers. Pictured in the capital Quito is* **Hawker Siddeley 748-267 Series 2A FAE-001** *(c/n 1684) of the Ecuadorian Air Force. It also has the civil registration HC-AUK and served to carry the nation's president. The interior was, however, not a total VIP fit, as it had normal airline seats for the bulk of the cabin with just eight extra-wide seats at the rear.*

Top: *The 748 had a special military version that featured a new rear fuselage shape incorporating a loading ramp that could be opened in flight, a taller tail fin, and – the most radical change – a kneeling undercarriage that brought down the tail loading bay height. It first flew in December 1963 and a total of twenty-one were built for the RAF. Most were later sold to civilian operators, especially those in Africa, but a small number have been retained for special needs. Pictured on take-off at Warton, Lancashire, is* **Hawker Siddeley Andover C.1 XS646** *(c/n Set No 30), operated by QinetiQ, then part of the Royal Aircraft Establishment. The different shape of the Andover is apparent in this photograph.*

Bottom: *Japan has not, to this day, built many airliners, yet it also produced a 'DC-3 Replacement'. NAMC (Nihon Aircraft Manufacturing Company) was formed in 1957 as a joint venture between a number of Japanese aircraft manufacturers, and the company's design was a conventional twin-engined turboprop with the Rolls-Royce Dart as the powerplant of choice. The first YS-11 flew in August 1962, and fewer than 200 were manufactured, but it did penetrate a number of markets including North and South America as well as Asian nations. Pictured at Komaki AFB, Nagoya, is* **NAMC YS-11P 02-1158** *(c/n 2150) operated by 403 Hikotai (Squadron) of the Japanese Air Self-Defence Force based at Miho airbase on the west coast of the country. As can be seen, the Japanese military is perhaps the most colourful of the world's air forces, as most nations paint their aircraft in various shades of grey.*

Below: *De Havilland Canada (now part of the Bombardier Group of companies) has had a long-deserved reputation for building aircraft that can be used in any location in the world and in any climate. The company's first civil turboprop was the DHC-6 Twin Otter, powered by a pair of Pratt & Whitney Canada PT-6A powerplants. It first flew in May 1965, and production continued until 1988, with more than 800 airframes being built. The demand for the type has been so strong that Viking Air of Sidney, British Columbia, has restarted production building completely new aircraft with a nineteen-seat capacity and STOL capability. The Twin Otter is destined to be seen all over the world for many years to come. Pictured at its Yellowknife, NWT, base is **DHC-6 Twin Otter 300 C-FASG** (c/n 373) operated by Air Tindi. This aircraft was built in September 1973 and is fitted with oversize, low-pressure 'tundra' tyres, enabling it to land on almost any surface it encounters in the far north of Canada.*

Top right: *Following the Twin Otter, de Havilland Canada produced the 'Dash Seven'. This was a high-wing, forty-eight-seater commuter powered by four PT-6A turboprops, and first flew in March 1975. It was renowned for its short-field performance and its ability to deliver a full passenger load to very small airports. It was also very quiet, and for a time the only aircraft that could operate into London City Airport in Docklands, as it fulfilled both the strict noise and steep glide slope requirements. Just 113 aircraft were produced, and now its successor, the much stretched DHC 8-400, can uplift seventy-eight passengers with twin-engined economy. Pictured at Yellowknife, NWT, is **DHC 7-103 Dash 7 C-FWZV** (c/n 81) of Air Tindi. This example was built in March 1982 and has a current seating capacity of forty-four.*

Middle right: *Built as a military transport to replace the earlier piston-powered Caribou, the de Havilland Canada Buffalo first flew in April 1964 with the power of a pair of 2,850shp General Electric T-64 turboprops. Very few aircraft are to be found with civil operators, despite it having the go-anywhere, carry-anything tradition of the manufacturer. However, one civil operator, Arctic Sunwest, is to be found at Yellowknife, NWT, and **DHC-5A Buffalo C-FASY** (c/n 107A) is one of a pair operated by the company; this one was built in 1982. The roles for this aircraft include taking stores and equipment to exploration camps and remote settlements. It has a payload capacity of 18,000lb and, with rear loading and a roller floor, speedy turnarounds can be achieved.*

Bottom right: *When Convair built its twin-engined 240/340/440 airliner range, they were all piston-powered. However, several other companies took on the task of fitting turboprops to existing airframes. One of the first was the installation of the British-built Napier Eland, such conversions being known as either Convair 540s or Eland Convairs. Rolls-Royce bought out Napier in 1962 and closed down the line, as it competed with its own Dart engine; aircraft fitted with Darts were known as Convair 600s. The most successful and long-lasting conversions have been those fitted with the Allison 501, and the designation for these is Convair 580. Pictured at Willow Run, Detroit, is **Convair 580 N51211** (c/n 489) operated by the Environmental Research Institute of Michigan. This organisation develops remote-sensing and technologies for geospatial information for both military and civil applications. This aircraft was built in 1957 as a model 440, and converted to a 580 in 1980.*

Top: *Kelowna Flightcraft now holds the manufacturing Type Certificate for all of the Convair twins, so it can manufacture any part required and is able to overhaul any fitted components. Both airframe and Allison engines can be maintained. Added to this is training for both pilots – the company has a simulator – and ground engineers. It has also produced a new stretched version (by 14ft 3in) with a cargo door on the port side of the rear fuselage. It has a modern glass cockpit, is powered by a pair of de-rated 4,300shp Allison 501 turboprops, and is known as the Convair 5800. Pictured at Vancouver, British Columbia, is* **Convair 580 C-GKFP** *(c/n 446), operated by Kelowna Flightcraft in cargo mode. This aircraft was first built as a model 440 in July 1957, first serving in South America before conversion to a 580 during 1969.*

Bottom: *In 1958 Short Brothers of Belfast, Northern Ireland, acquired the rights of a Miles Aircraft freighter. What followed was almost a completely new design with a square box-like fuselage, high wing and twin fins. This aircraft first flew in January 1963 and was powered by a pair of 390hp Continental piston engines, which were soon replaced by the Turboméca Astazou turboprops in October of the same year. Production continued until 1989, by which time 154 examples had been built and the type still serves in many parts of the world. Pictured at Yellowknife, NWT, is* **Short SC-7 Skyvan 3 C-GKOA** *(c/n SH1905) in the livery of Summit Air. This 1972-built aircraft is one of three operated by the company and is fitted with nine seats, the other two being configured for freight and utility roles.*

Top: *The Skyvan has many roles, from cargo to passenger; and is also operated in Florida as a tanker, spraying insecticide to kill mosquitoes breeding in still-water areas. The military use of the type varies from cargo-carrying to parachute-training, as the rear opening doors make it a perfect jump ship. Pictured at Payerne, Switzerland, is* **Short SC-7-3M Skyvan 400 5S-TA** *(c/n SH1855) of the Austrian Air Force. This 1969-built aircraft has been operated on both wheels and skis during the winter period.*

Bottom: *In the west the two medium four-engine turboprops, the Bristol Britannia and the Lockheed Electra, had relatively short lives with the major airlines of the world on their prime routes. This was because the paying passengers wanted pure jets, and got them. In Russia and the other Soviet bloc countries 'passenger choice' was not a phrase to be heard – you got whatever aircraft was on the route, and if you didn't like it, the only option was not to travel. This mind-set gave the Ilyushin IL-18 a far longer front-line service life than its western counterparts. The type first flew in July 1957, and most models were powered by four 4,250shp Ivchenko Al-20M turboprops; seating was up to 100 in a high-density configuration. Pictured at Moscow's Domodedovo Airport is 1967-built* **Ilyushin IL-18D RA-75454** *(c/n 187010104) of Rossia (Russia State Transport Co).*

Top left: *To replace the piston-powered Ilyushin IL-14 the Kiev-based Antonov company produced the first of what would be a very long line of airliners and their developments. The An-24 first flew in December 1959 under the power of a pair of Ivchenko AI-24 turboprops. It was a conventional high-winged aircraft, but able to handle the many short and rough fields in the heartland of Russia. Pictured on the ramp at Moscow-Bykovo Airport is **Antonov An-24ALK RA-48395** (c/n 07306209); built in August 1970, it is one of four flown by the company, and is operated by Flight Inspections & Systems in the role of calibrating airport navigation aids.*

Middle left: *Pictured at Moscow's Vnukovo Airport is **Antonov An-24B RA-47289** (c/n 07306509), built in 1970 and configured for passengers with forty-eight seats in a single class. It is owned by UTair Aviation based at Khanty-Mansisk, which operates a very mixed fleet from small to large helicopters and, in fixed-wing, from utility biplanes to tri-engined jet liners.*

Bottom left: *The cargo version of the An-24 had a redesigned rear fuselage; its loading ramp has a mechanism that can be opened in flight to drop supplies or parachutist, the fuselage windows have been reduced in number and the floor hardened for load-bearing. Pictured at Berlin-Schönefeld is 1972-built **Antonov An-26 1406** (c/n 1406) operated by Krakow-based 13 PLT (Transport Aviation Regiment) of the Polish Air Force. On the opposite side is a large bulged observation window just aft of the cockpit.*

Above: *Another aircraft developed from the base-line An-24 is the specialised photographic and mapping survey variant, the An-30. It is instantly recognisable by its glazed nose and belly doors for vertical camera mounting. Pictured on the ramp at Moscow-Bykovo Airport is **Antonov An-30 RA-30075** (c/n 1306) operated by Myachkovo Air Services. The type has been used by a number of air forces from the original Soviet bloc to operate 'Open Skies' flights over western military establishments as part of an East-West agreement. Some air forces even paint 'Open Skies' titles on their aircraft.*

A view that says it all about the lure of the propliner: that tramp streamer of the skies, the Curtiss C-46, is flying along in Colombia with the emergency exits removed so you can lean out and take pictures. It could only happen in the world of pounding pistons, sadly one that may not last for many more years.